Ground Penetrating Radar

Ground Penetrating Radar
Theory and Practice

Erica Carrick Utsi

EMC Radar Consulting, Ely, United Kingdom

Butterworth-Heinemann
An imprint of Elsevier

Butterworth-Heinemann is an imprint of Elsevier
The Boulevard, Langford Lane, Kidlington, Oxford OX5 1GB, United Kingdom
50 Hampshire Street, 5th Floor, Cambridge, MA 02139, United States

Notices
Knowledge and best practice in this field are constantly changing. As new research and experience broaden
our understanding, changes in research methods, professional practices, or medical treatment may become
necessary.

Practitioners and researchers must always rely on their own experience and knowledge in evaluating and
using any information, methods, compounds, or experiments described herein. In using such information or
methods they should be mindful of their own safety and the safety of others, including parties for whom they
have a professional responsibility.

To the fullest extent of the law, neither the Publisher nor the authors, contributors, or editors, assume any
liability for any injury and/or damage to persons or property as a matter of products liability, negligence or
otherwise, or from any use or operation of any methods, products, instructions, or ideas contained in the
material herein.

British Library Cataloguing-in-Publication Data
A catalogue record for this book is available from the British Library

Library of Congress Cataloging-in-Publication Data
A catalog record for this book is available from the Library of Congress

ISBN: 978-0-08-102216-0

For Information on all Butterworth-Heinemann publications
visit our website at https://www.elsevier.com/books-and-journals

Working together
to grow libraries in
developing countries

www.elsevier.com • www.bookaid.org

Publisher: Joe Hayton
Acquisition Editor: Ken McCombs
Editorial Project Manager: Katie Chan
Production Project Manager: Julie-Ann Stansfield
Designer: Matthew Limbert

Typeset by MPS Limited, Chennai, India

Contents

Acknowledgments

The author would like to thank the following colleagues for generously sharing information and thoughts on current developments in the GPR market.

Norman Bell, Allied Associates, European agent and distributor for GSSI
Peter Annan, Sensors and Software
Guido Manacorda, IDS Ingegneria Dei Sistemi SpA
Alan Jones, Pipehawk plc
Reinaldo Alvarez-Cabrera, Geoscanners
PAV Utsi, Utsi Electronics.

Her thanks are also due to Guido Manacorda in his capacity as Licensing Officer of EuroGPR for advice regarding the current European licensing regime and to Peter Annan for similar advice regarding the Canadian licensing regime.

The author would like to thank both Mike Forde, Emeritus Professor of Civil Engineering, University of Edinburgh, and Dr. Richard Chignell, formerly of Pipehawk plc for their support and encouragement for this project. She would also like to thank Dr. Richard Chignell and Matthew Guy, Geophysicist at Geomatrix Ltd. for their helpful comments on sections of the text. Any errors are, of course, the author's own.

She would also like to thank the following people and organizations for allowing GPR-related material to be used as illustration.

Alex Birtwisle of Atlas Geophysical Ltd. for permission to use Fig. 1.1.

Dr. Richard Jones of Glasgow University for permission to use the data in Fig. 4.1.

The late Jerry Hamer of Glasgow University Archaeological Research Division for permission to use Fig. 7.2, previously published in Utsi, E., 2007. Wetlands viewed through the antennas of a Ground Penetrating Radar. In: Barber, J. et al. (Eds.), Archaeology From the Wetlands: Recent Perspectives. Proceedings of the 11th WARP Conference. Society of Antiquaries, Edinburgh, pp. 213–220.

PAV Utsi for generating Fig. 2.1, as an illustration of a frequency range.

Utsi Electronics Ltd. for permission to use all data from surveys carried out by the author during her time working for the company.

Introduction

Ground Penetrating (or Probing) Radar, more commonly known as GPR, is simply one of the most useful geophysical techniques available to engineers and scientists. It allows us to detect what lies beneath the ground or within buildings. It is primarily a safety aide since it is used to prevent potentially dangerous situations arising or for the detection and treatment of hidden problems. As such, it has a wide remit whether this is the accurate location of pipes and cables in order to prevent damage during site development, the detection of crevasses in glaciers or the internal structural investigation of roads, bridges, airports, and historic buildings to detect problems before they become critical.

This book is an introduction to GPR. It is intended to be a practical guide, used as a basic tool by survey practitioners and also by those who would find GPR data helpful but generally commission someone else to do the work. For both groups a basic understanding of how GPR works, what can be achieved with it, the decisions that need to be taken before a GPR survey is begun, and the potential problems (and solutions) is vital. There is a view that GPR is unreliable and therefore not always worth the cost of using it. The aim of this book is to show how some of this apparent unpredictability can be removed.

Although there can be a strongly mathematical component to GPR, advanced mathematical ability is not required for this introduction. A certain amount of calculation is involved but none of this is more complicated than basic relationships such as speed = distance/time. The reward should be better survey decisions, better survey results, and a clearer understanding of what the risks of failure are.

Who uses GPR? A growing number of professionals including civil and structural engineers, surveyors, and those concerned with managing a wide variety of environmental resources and archeologists. The ultimate end user of the information generated is often government, local, or national, and the primary reason for use is very often the safety of individuals and the general public.

The technology is essentially the same as traditional radar. The big difference is that instead of probing into air in search of large metal objects such as planes or cars, GPR aims to uncover buried targets, in the ground, within buildings, essentially anywhere where they cannot be seen but someone needs to know that they exist and/or what state they are in. The main difference between the more traditional radar used in air traffic control, for example, and GPR lies in the parameters of frequency and wavelength.

Since radars measure depths and targets in fractions of their wavelengths, this means that a single radar system cannot fit all applications. Since GPR can be used for a wide variety of applications from the shallow depths and small targets of security scanning to the far greater depths of glaciology or mining, GPR systems also differ considerably in their characteristics and uses. One of the biggest current misuses of GPR is the application of a radar to detection for which it is not suitable, often, for example, using the wrong frequency of antenna. An unfortunate side effect of this is that GPR technology may be inappropriately blamed when the real reason for failure is the use of the wrong GPR equipment for the task in hand. A clear grasp of how GPR works and the difference between using the wrong equipment (which can, at least, be corrected) and the real limitations of the technique is absolutely essential both to the GPR operator and to the client requiring the information from the GPR survey.

It is difficult currently for someone starting out using a GPR to find a basic guide to the technology. There are many academic books dealing with the physical principles lying behind radar, including GPR, not least of which is the seminal work by David Daniels. Others deal in detail with the intricacies of Maxwell's equations, and different methods of processing and presenting data. There are also good texts on particular GPR applications. However all of these books are written with the assumption of a certain level of knowledge and understanding of physics that can be a deterrent to both the beginner and the person whose expertise lies in an associated field. This book aims to bridge that gap. There is a reference list at the end of each chapter and a further reading list in the appendices for those who wish to pursue the subject of GPR beyond the simple basics and also for those who wish to extend their knowledge of particular aspects of the technology.

There are a number of current problems within the world of GPR which need to be addressed urgently. The first of these is that although many end users of the technology are very competent and professional in their use of the technology, there are others who do not understand the underlying principles as fully as they might, a situation which is not helped by the technology being seen as "difficult." The lack of a basic understanding of the principles and applications of GPR, of what information can and cannot be derived from a GPR survey and the preparation which may be necessary can easily lead to false expectations. This makes the potential GPR client vulnerable to inappropriate claims, either from those whose enthusiasm outstrips their technical knowledge or from false claims of inappropriate variations of the technology. Unfortunately, because many geophysical techniques, GPR included, are seen as a "black art" with poorly understood limitations (outside of the GPR community), there may always be a temptation for the unscrupulous to make money at the expense of those who need the technology by promising to overcome the physical laws of the universe as we know it, relying ruthlessly on a perceived lack of scientific knowledge of their client group.

Such confusion may arise through lack of discussion between the client and the GPR surveyor. If these discussions are to be meaningful, the client needs a basic understanding of the technical issues involved and an appreciation of what issues should be addressed before the survey is commissioned. Without this basic level of understanding it can be very difficult for potential GPR clients to judge what they actually require from a GPR survey or what information they should give in advance to the surveyor. The net result risks being a mismatch in the expectations of the client and the outcome of the GPR survey. Once fingers have been burned by receiving poor or no results, the reaction of the client may be to avoid the technology in the future. Although this is not commonplace, there have been instances where projects that could usefully have used GPR have not done so because not only is it a "difficult" technology but it "probably will not work." This is a great pity because, even with its limitations, GPR is an extremely useful technology which can provide useful, interesting, and often lifesaving information. The limitations in practice are not always those of the technology. They may have their origin in misunderstanding. A primary objective of this book is to provide a simple guide to facilitate meaningful interchanges.

Although training in GPR is widely and professionally carried out by all manufacturers, this tends to apply only to their own systems, rather than to the wider technology. There is, quite rightly, an emphasis on the mechanics of operation rather than the more general principles of the technology. On the other hand, training in specific applications focuses primarily on the particular topic without necessarily going into detail of the underlying principles. It is not uncommon for this type of industry training to treat GPR as the difficult younger sibling of other technologies. This may limit the degree to which the technology is explained and understood.

If we take the example of utility detection, GPR is often seen as the poor relation of electromagnetic location (EML) as the combination of "cat" (cable avoidance tool) and "genny" (signal generator) is relatively inexpensive, simple to use and in principle capable of detecting all metal services. The danger is that there are now many plastic utilities installed in addition to other nonmetallic structures present in the subsurface, gas pipes being an obvious example. These require GPR for their detection. The principles of using EML are widely discussed in utility detection training and there is a general understanding both of how the equipment should be used and what can lead to mistakes. The application of GPR, on the other hand, is not always as comprehensively discussed.

There is also disagreement within the industry over the extent to which it is appropriate to apply the technique. The problem is frequently made worse by competitive pricing and acceptance of the lowest cost tenders. This can only be acceptable if there is a clear understanding of the potential risk of failing to detect targets, understood by both the client and the GPR surveyor. In spite of efforts within the utility detection industry to clarify different levels of survey and what each of these might mean, there is still a tendency to deploy GPR as though it were able to detect optically with all targets equally visible. Since this is not how GPR works, sooner or later, targets will be missed, adding to the illusion that the technology is unreliable. There have been occasions when a client has specifically requested that GPR should not be used due to the cost and to the unlikelihood of useful results. This tendency to restrict the use of GPR unfortunately is the exact opposite of what is required for a successful survey.

The last strand of problems derives directly from the variable levels of training. There are times when a GPR survey will not give the hoped for result. Where this happens, it is not unusual for little or no explanation to be given as to why the survey was unsuccessful. The reasons for failure are not always down to the radar. Although it is always easier to blame the technology, sometimes the problems lay at the planning stage, e.g., using the wrong frequency of antenna, or in the lack of understanding of the user, the effect of groundwater being a good example (see Chapter 4: The Effect of Water and Air).

This book is intended to provide a basic understanding and guidance to those who are either already using the technology or wish to begin using the technology. It is also intended for those professionals who could benefit from obtaining GPR-based information but whose background does not lie in geophysical techniques and who therefore wish to commission a survey rather than carry one out themselves. The aim is to explain, in straightforward and relatively simple fashion, what is important in using a GPR and why; how to take the critical decisions of which frequency of antenna to select and what sort of survey coverage will be required; and to make a reasonable judgment as to whether GPR is a suitable technique for the planned project. It should also allow those who commission GPR surveys to take a judgment as to whether an apparent failure of the GPR survey is due to the unsuitability of the site or one or more poor decisions by the surveyor. Guidance is given both on legitimate reasons for failure and excuses which are unacceptable and may therefore be due to human error rather than the technology. Although GPR will always seem to be an alien technology (for reasons explained in Chapter 1: Fundamentals of GPR operation), there is no reason for it to be seen as a difficult one nor for it to be avoided. It is much too useful a tool.

The material in this book was developed from international training courses provided by the author to clients, manufacturers' agents, government and European agencies, engineers, land surveyors, archeologists, and university students, including material provided as part of the European COST action TU1208, Civil Engineering Applications of Ground Penetrating Radar. The purpose of these courses ranged from specific GPR applications to a more general understanding of GPR technology.

In addition to describing how GPR works and the basis on which it should be used, practical examples, based on real survey assignments, are given to allow the reader to test his/her understanding of the principles (Chapter 12: Examples of Practical Problems). There is also a basic GPR survey assessment checklist (Chapter 7: Survey Strategies), guidance on how to access reputable equipment as well as finding a GPR surveyor who has committed to good practice, and a further reading list (Appendix A: Further Reading and Appendix B: Professional Association and Reputable Manufacturers) for those who wish to engage further with the technology.

The author is a former director of a GPR manufacturing company, a former Chairman of the European GPR Association and a very experienced GPR surveyor with specializations in the innovative application of GPR to engineering problems, archeological and wetland exploration, as well as the provision of technical GPR advice to industry.

Chapter 1

Fundamentals of GPR Operation

Chapter Outline

THE "DIFFICULTY" UNDERLYING GPR: VISUALIZATION

The main problem in understanding Ground Probing Radar (GPR) data is that, as people, we rely heavily on the evidence of our eyes throughout our daily lives. Without needing to think about it, we read, recognize people, places, and objects using optical techniques. We are therefore predisposed to think in optical terms, regardless of the technology which we happen to be using. As part of this process we expect to see objects regardless of their size or their proximity to other material. GPR relies on the transmission of electromagnetic pulses (radio waves) and does not present data automatically in optical form. Although we can generate images from the GPR data, it is essential to understand that these may not mean quite what we expect them to. The radar is always right. Our interpretation may or may not be correct.

Ideally every user of GPR data would like to be presented with some form of picture of the targets that they wish to see. The structural engineer who needs to know the size, distribution, and volume of voids in a concrete lined tunnel expects an image of those voids, preferably one which allows him/her to calculate the amount of filler material directly. The construction company commissioning a utility survey needs a plan of existing pipes and services on their development site. Environmental specialists, archeologists, and mining engineers would be extremely happy with output which outlines the different stratigraphic layers and the features contained within them. Unfortunately, neither we nor GPRs can "see" into the ground in that sense. To use GPR successfully, we need to understand the interaction of radio waves with the ground or other survey media such as concrete, bricks, and

Ground Penetrating Radar. DOI: http://dx.doi.org/10.1016/B978-0-08-102216-0.00001-1

other materials used in construction which may require investigation by GPR. In each case, we need an understanding of why wavelengths matter and what exactly it is that the radar is responding to in each environment in which we use it.

For many of the same technical reasons, there is uncertainty over the capability of any individual radar antenna in surveying any given site unless the technical considerations can be addressed in advance. Although there are many very competent and professional GPR users around, there are also many who are uncertain as to how to decide which antenna or which system should be used or who make the assumption that the radar they have immediate access to will be capable of carrying out virtually any task. The issue is not which of the many radar systems to use, rather it is a question of understanding that our own built-in method of detection and that of the radar are based on different principles. There is a translation job to be done both before and after the survey. Rather than taking blanket decisions such as "GPR should always be used to detect subsurface cracking beneath flexible pavement" or "A 500 MHz antenna can be used down to a maximum depth of X m (or X feet)," it is better to analyze the situation in so far as the relevant information is available. The process is not complicated and, if done properly, will definitely improve detection rates. Although both of the statements quoted here are potentially true, they are not universal, i.e., there are sites and circumstances where they might not apply. Understanding why this is, is the key to better surveying.

In trying to make the technology accessible to a wider pool of users, many manufacturers have tried to simplify the process of selecting survey parameters, for example, by having "shallow" and "deep" depth settings rather than asking the user to work out the depth to which he/she wishes to survey in nanoseconds time. The initial purpose behind these design features is to simplify certain types of GPR survey, usually either utility detection or structural investigations, and for these specific purposes this can be very useful. However, it is not uncommon to find these same radars applied to other types of survey task. Although using this terminology has undoubtedly made GPR more attractive to greater numbers of people and will have helped with many surveys, it also provides a major limitation on the use of the technology. The first obvious problem is that there is no general agreement between manufacturers as to what simplifications should be built in and the variety of shortcuts built into a system are often specific to that one system. So, for example, where several manufacturers build in settings of "shallow" and "deep," what constitutes "shallow" or "deep" varies from one system to another. Worse, from the point of the GPR user, this also depends on the use being made of the radar. Shallow in terms of utility detection is not necessarily the same as "shallow" for a road investigation and very unlikely to bear any relation to the same term in an environmental survey. The common understanding of these terms will also vary from country to country. These are relative terms and it is useful to understand the basics of GPR that underlie them.

In essence, the manufacturers are taking their best guess at the user's site conditions. This is very useful provided that those approximations fit the site to be surveyed. The main problem is the risk that this discourages the user from challenging their own assumptions about the site they are planning to survey rather than reverting to the fundamental principles of GPR. Since there will always be sites that do not fit the preconceptions or presumed averages of the equipment designer, the operator may not be able to make the appropriate decisions to alter the parameters of his/her survey to suit the reality of the site. This may be a question of how to change the settings on the radar but could also be making sure that the radar actually has the requisite features and capacity to tackle the survey at all.

It is therefore very important to have a basic understanding of GPR theory in order to be able to use the equipment effectively.

HOW GPR WORKS

Every GPR is made up of a control system, at least one transmitter antenna, at least one receiver antenna, and some means of data storage (Fig. 1.1). A transmitter antenna and receiver antenna together is termed a transducer pair. It is also usual to have a viewing system for monitoring the data in real

FIGURE 1.1 The basic elements of a GPR system: 1, controller; 2, transmitter antenna; 3, receiver antenna; 4 and 5, laptop for data storage and monitoring; and 6, encoder wheel for distance measurement.

time, either a laptop, a notebook or other computing device, a data logger, or a similar type of screen if the computing device is built in. Many systems also contain a distance measurement device, typically an odometer (or encoder) wheel. Some systems use GPS or total station instead of a wheel and others can be used with both.

The GPR controller generates electromagnetic pulses which are passed into a transmitter which then transmits into the survey medium (the soil or other material under investigation, e.g., concrete). As the signal passes through this material, parts of the signal are returned to the receiver. It is the changing nature of the buried material in the subsurface that triggers each of these reflections. The returned signals carry information back through the receiver to the controller and from there to the laptop, tablet, or other viewing device. This information is not optical and the radar does not give a picture of what lies in the subsurface, at least not in the traditional meaning of the word.

The survey medium depends on the environment under investigation. For GPR this is most commonly the ground, either open soil or man-made substances such as asphalt and concrete. It can also mean walls, ceilings and floors of buildings, ice, and any other substance within which lie targets to be investigated. From this point in the text on, where reference is made to GPR being used on the ground, these other environments are included in the definition. "Ground" is used only for simplicity.

The basis on which parts of the signal are returned is the difference in electromagnetic properties between one substance and the next. In simple terms this means that the radar can detect any material provided that its magnetic and/or electrical responses are different from the other materials which surround it. The measure of difference is the material's relative permittivity (sometimes known as dielectric constant) for which the commonly used symbol is Er. One of the major advantages of GPR is that it responds to any material and is not limited, for example, to metal detection. Provided that the material for which a search is being made is different in electrical and magnetic response to its surrounding environment, it can potentially be detected by a GPR. So, for example, it is perfectly possible to detect buried plastic pipes by using a GPR.

Laptops, data loggers, tablets, and PCs are used to set the controller's survey parameters (e.g., probing depth, frequency of sampling, number of antennas) and also the on-screen display settings which are for the benefit of the person viewing the data as these are collected. The computing device is also used to store the data. Some systems record the viewing parameters so that the data are effectively processed as they are collected, others allow processed data to be viewed in real time but store raw data for postprocessing. The computing device may be built into the radar or may be separable, it depends on what the manufacturer perceives as the preference of the majority of their clients.

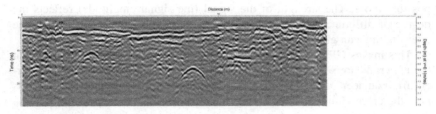

FIGURE 1.2   An example of GPR utility detection data.

One thing it is worth being wary of is download of the data as an image only. Although this does provide a record of the survey, it means that the data can never be interrogated by alternative postprocessing. If the processes applied during survey were inadequate in anyway, there is nothing further that can be done with that data. This is not true of systems which record raw or processed data. As long as the data can be either downloaded or transferred into a suitable interpretative package, it is possible to challenge the assumptions made at the time of the survey and to apply other processes to the data in order to make sure that the best possible level of detection has been achieved.

There are many different designs of GPR and GPR antennas. This is partly because GPR is used in many different applications for which different mechanical and electronic designs are appropriate. Most of the common ones are described in Chapter 9, Antenna and Screening.

As the GPR is moved across the survey area, the sequence of signals builds up a two-dimensional image of the subsurface in a plot where the x-axis represents the distance traveled by the radar (typically measured by an odometer) and the y-axis represents the depth probed and from which each signal has been returned (Fig. 1.2).

WHAT DO THE SIGNALS REPRESENT?

The first thing to notice about this two-dimensional data is that, although the image represents a vertical view of what lies in the ground, the resulting image in no way resembles the shape and form of the actual objects below ground (Fig. 1.2). Much as the user might like to have an optical image, this is a radio wave representation and it does not and cannot look like the objects would if the user were able to use a camera.

One of the reasons for this is that each signal does not represent the target itself. The signal is generated by the boundary between two or more objects (or two or more features). Signals are reflected when the electromagnetic properties of the subsurface materials change. If the electromagnetic properties do not change, no signal will be returned to the radar receiver. So every signal is the product of a change from one material to at least one other, possibly more. The strength of the signal (the signal amplitude) reflects the degree of difference between these two or more substances. As we will see later on, this can give very useful information.

This makes GPR a very powerful technique since it does not matter what the materials are so long as their electrical or magnetic properties are different from adjacent objects or their surrounding environment. GPR detection is not, therefore, limited to detecting metallic objects. It is possible for the radar not to detect targets but only if the electrical and magnetic properties of these targets is very similar to the environment that surrounds them.

The next thing to observe is the distinctive shape of the signals in the data. These are primarily either hyperbolas or lines. In both cases, they are outlined by a short series of black and white bands. Although an inexperienced viewer might be tempted into thinking the hyperbolas represent the outline of the pipes and cables buried in the subsurface, this is not normally the case. An object does not have to be cylindrical to be the source of a hyperbolic signal return.

HOW HYPERBOLAS AND LINES ARE FORMED

A radar beam is not linear, it forms a cone shape beginning at the transmitter antenna and disseminating outward from there. For GPR this normally means into the ground. There is an optical way of illustrating this. Take a torch and shine it on the ground or on a wall close to you. Then raise the torch higher from the ground or back away from the wall. The further you move away from the ground or the wall, the wider the footprint of the beam becomes. The radar beam is similarly formed although its footprint is oval in contrast to the circular footprint of a torch. The closer the target is to the ground surface, the narrower the footprint of the beam. The deeper the radar probes, the wider the beam. It follows, therefore, that the radar will start to detect buried objects before it reaches the point directly above them.

As the radar approaches the position of the target, the edge of the beam will begin to detect the target. The radar measures the distance from the antennas to the target as its apparent depth (Fig. 1.3). However, this distance, being at an angle to rather than directly above the target, is longer than the real depth of the target which would normally be measured vertically. As the radar approaches the target the apparent depth (or distance to the target) shortens until it reaches its minimum directly above the position of the target. The apparent depth will then increase again as the radar leaves the target behind.

If we were to plot the depth of the target against the distance traveled by the GPR, this would form a hyperbolic shape. Fig. 1.4 illustrates this. The curve is formed by the size and shape of the radar beam footprint starting at the point at which it first detects the object, continuing through to the position where the radar lies directly above the object and then on

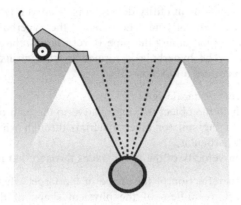

FIGURE 1.3   Diagram of GPR beam detecting a target.

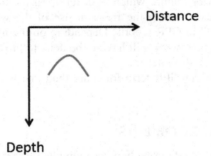

FIGURE 1.4   The plot of distance vs depth results in a hyperbolic shape.

through to the last point at which the beam clips the object before contact with the target is lost. The hyperbola outlined in the data is formed by this travel path (cf Figs. 1.3 and 1.4). In reality the apparent depth is measured by a cone-shaped beam, of course, not a vertical line but the principle remains the same.

In simple depth terms the radar is closest to the object when the midpoint of the antenna lies directly above the target. Before the radar reaches this point, the detection path is not vertical but follows a series of angles (cf Fig. 1.3). For the whole of this travel path up until the receiver lies directly above the target, the distance between the receiver and the object will be greater than the actual depth of the object. As the radar approaches the point directly above the object, this distance shortens. As the radar passes over the object and continues beyond it, the apparent depth will increase again.

It is important to understand this concept because it has implications for understanding the data generated by the GPR. For example, one of the most commonly asked questions in utility detection is "Can you tell the size of the pipe?" The answer is generally no. The shape of the hyperbola is not formed by the outline of the pipe unless the pipe is very large relative to the wavelength of the radar. The hyperbola is formed by the intrinsic shape of the radar beam. It is therefore a function of:

- the manufacturer's radar design;
- the angle at which the object is lying relative to the path of the radar;
- the properties of the soil (or other material) through which the beam is passing and, in particular;
- the transmission velocity of the radio waves through that material.

It is not a direct reflection of the shape of the target whether it is a pipe or any other object, regardless of the physical shape of that object. The hyperbola is simply the radar's depiction of any target in the subsurface. Lesson number one is therefore that a hyperbola represents a feature in the ground but it may not be the target we are searching for. The hyperbola represents an object or feature which is different in its material composition from what surrounds it, hence the frequent use of the words "anomaly" or "anomalous material" in GPR reports. Depending on the nature of the survey, it is probable that further work will have to be done to prove that this anomaly is the target that we wish to detect.

Note that the correct GPR term for more than one hyperbola is hyperbolas and not hyperbolae.

LAYERS AND LARGE OBJECTS

The previous paragraphs describe how an individual target will be depicted in GPR data. The outline of a layer in the ground is formed in precisely the same way as the simple case just considered of a single target. Since layers are generally continuous, however, a continuous series of hyperbolas will be formed as the radar travels across the layer change. These hyperbolas join up to form a continuous line which follows the outline of the boundary between the layer containing one type of material (or group of materials) and the next material or set of materials with even a slightly different composition (Fig. 1.5).

A discrete flat object buried in the subsurface will appear slightly differently. The majority of its length will be reflected as a continuous line in the GPR data. At each end, however, there is no other signal to mask the tails of the first and last hyperbolas so each side will appear to curve downward rather than ending abruptly. This is the case even if the object is lying at an angle relative to the ground surface. It is a function of the numerous reflected hyperbolas forming the line, only the first and last of these will be fully visible. Care needs to be taken in interpreting this sort of short line in GPR data. Does it represent something lying directly below or is it simply the corner of a flat object?

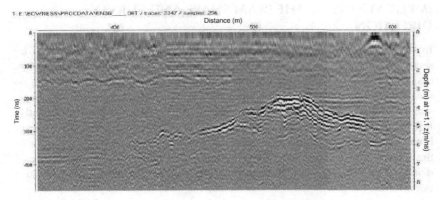

FIGURE 1.5 Multiple stratigraphic layers from an environmental GPR investigation.

Fig. 1.5 shows the outline of a number of subsurface layers taken from an environmental GPR investigation. The continuity of the layers is maintained even where their depth below the ground surface changes.

The leading and falling edges of each hyperbola are not visible because they are masked by the returns from the hyperbolas on either side. The resulting picture is very much easier to understand and interpret than an individual hyperbola because it resembles what we would recognize in a section drawing.

We have seen how the radar depicts small targets as hyperbolas. For large objects it may be possible to trace part of the target's outline in exactly the same way that the radar would trace the boundary of a particular layer. In practice this means that the object has to be large relative to the size of the wavelengths emitted by the radar.

TARGET LOCATION

For discrete objects, it follows from the way in which hyperbolas are formed, that it is the shortest distance, the top of the hyperbola, which indicates the position of the target in the ground. In two-dimensional data such as Figs. 1.2 and 1.5, this allows both a depth reading from the vertical y-axis and a distance measurement from the start of the survey along the x-axis. It follows that the start position needs to be recorded independently of the radar data unless a GPS is being used simultaneously with the GPR.

Although most GPRs enclose both transmitter and receiver antennas within a single box, it is pertinent to ask at what point does the GPR lie directly above the buried object? The traditional measurement point on the antenna is the midpoint between transmitter and receiver. Most commercial GPRs have an external marker, usually on the antenna casing, indicating where this point lies.

IMPLICATIONS OF THE BEAM SHAPE AND SURVEY DIRECTION

In the torch experiment described above, the footprint of the beam can be seen to be circular, i.e., it is symmetrical in all directions. Optical perception does not change with the orientation of the detection equipment except in so far as the shape of the object changes. If you move around a room, the basic shapes of the objects and people in that room do not change significantly. For radar this is not the case. The shape of the radar beam is not the same in both orthogonal directions. Instead of having a circular footprint, the base of the beam is oval. This means that, unlike optical methods, the probability of detection is not equal in all directions and survey direction is, as a result, important.

If a GPR crosses a target, e.g., a pipe buried in the ground, this information will show in the data as a hyperbola in the manner outlined above for discrete targets. The best possibility of detection is if the radar crosses the feature at right angles (90 degrees). Detection of the target is possible at any angle provided that the radar crosses the pipe, rather than traveling along it. The optimum direction is for the radar to cross the target at a right angle, because the electrical field (e-Field) of the radar lies orthogonally to the orientation of the radar's antennas. At an angle of 90 degrees, the pipe is fully aligned with the radar's e-Field and at this position its likelihood of being detected is greatest.

If the GPR travels along the same direction as the pipe, it will either not be seen at all or it may appear as a layer change because of the continuity of signals along the length of the pipe. The outcome depends upon the size of the pipe. If the pipe is small, it is likely that it will not be detected by the radar. If the pipe is large, it will appear as a continuous line, effectively mimicking a layer change in the ground. Only if a parallel survey line is carried out nearby, one which does not detect this pipe, will it be apparent that this result indicates an object rather than a change in the overall ground conditions.

The implication is that survey in only one direction will not necessarily detect all targets. For this reason, it is not unusual to carry out two surveys over the same area, one at right angles to the other. If the buried targets are large, e.g., substantial wall footings, this may not be necessary, particularly if a closely spaced area survey is carried out. For small targets or where the radar is used only to sample, there is a major risk of nondetection. Although the technology is often blamed for failure to detect, the fault often lies in the understanding of the radar operator and his/her use of the radar on-site. Where a GPR search for small targets is required, it is essential to search in two orthogonal directions.

This assumes that both antennas are aligned along the same polarization which is the case for the majority of commercially available antennas. It is, however, also possible to survey using antennas which have been aligned at right angles to each other. These are known as crossed dipole antennas. They are generally used for small linear targets, typically joints in concrete, cracking, or small buried plastic mines (cf Daniels, 2004). There are crossed dipole antenna systems available commercially although there are fewer of these.

THE DOUBLE Y-AXIS

Looking at Figs. 1.2 and 1.5, there are three axes, not two, visible. The x-axis represents the distance traveled by the radar and the two y-axes the probing depth of the radar. By convention, the axis on the left is the depth measured in nanoseconds time. Radars measure in time rather than distance because radio waves do not travel at a constant speed. The transmission velocity of the electromagnetic pulses varies with the electrical and magnetic properties of the material they are passing through. The right-hand axis is the translation of depth into centimeters/meters or inches/feet using a calibrated velocity of transmission. Transmission velocity is discussed more fully in Chapter 4, The Effect of Water and Air.

CAUTIONARY NOTE

It should be evident from the contents of this chapter that technology which claims to "see underground" and produce the outline of objects in the subsurface should be treated with great caution. "Seeing" is not a normal feature of GPR although the use of the word may simply be unfortunate. The information provided by the radar is invariably correct but our understanding of this information may be limited because, even with experience, we do not naturally think in terms of changing electromagnetic properties as easily as we recognize color and shape optically.

This is not to say that we cannot produce some form of visualization from the GPR data but, as for many other new technologies that allow us to explore space or the deep ocean floor, this requires the input of additional computer processing, based on our interpretations of the original data. The demanding feature of GPR is that the transmission speed of the signal through the ground and the depths to which signals reasonably travel are both highly variable and identification of specific materials may be problematic. Both use and interpretation need a basic understanding of how the data are formed.

REFERENCE

Daniels, D.J., 2004. Ground Penetrating Radar, second ed. (Chapter 2).

Chapter 2

Wavelengths and Why They Matter

Chapter Outline

In this chapter we will consider why antenna frequency and wavelengths matter and why they may pose a problem for a survey if not thought about at the planning stage. It is unfortunately quite common to use a particular frequency of antenna because that is the equipment which is available without considering whether it is the appropriate tool for the job. Although there is no reason why GPR data should not accurately reflect the position of its targets, radars cannot measure with pinpoint precision. Radars, whether ground penetrating or air scanning, measure in units dictated by the wavelength of the radio waves they emit.

FREQUENCY AND WAVELENGTH

Whatever radar is used, it is critical to understand the implication of the wavelengths being transmitted. The wavelength determines:

- the maximum depth of penetration;
- the size of the smallest object which can be detected;
- the spacing with which adjacent objects can be separately detected; including
- whether or not the radar can be used on reinforced concrete; and
- the accuracy with which we can measure the depth of our targets.

Ground Penetrating Radar. DOI: http://dx.doi.org/10.1016/B978-0-08-102216-0.00002-3

In practical terms, therefore, being able to calculate the wavelength means the difference between being able to:

- ensure you have the right antenna for the detection job;
- ensure you have the right antenna for the size of your target(s);
- penetrate reinforced concrete;
- operate in wet conditions;
- select the right antenna to distinguish adjacent targets;
- provide an estimate of the accuracy of your work

or blaming the technology (possibly inappropriately) for an inability to do some or all of the above.

A further complication is that wavelength is not a constant, even for the same frequency of antenna. Wavelengths depend on the frequency of the radar but also on the material under investigation. Specifically, this means that the velocity at which the radio waves can pass through that material is also a factor. The basic equation for the calculation of wavelength is:

$$\lambda = \frac{v}{f}$$

where λ is the wavelength (in meters), v is the transmission velocity of the radio waves transmitted by the radar (in meters per nanosecond), and f is the frequency of the radar (in GHz). This is a simple equation which can be very useful to know and understand when working with a radar in different environments.

PEAK ENERGY

The frequency of the radar is effectively a constant, e.g., a 400 or 500 MHz system. Although in reality every radar emits a range of frequencies and the transmission of those frequencies also depends in part on the electromagnetic response of the ground, the frequency designation generally indicates the frequency at which the antenna transmits its maximum energy. The normal designation of an antenna is in terms of their "mid-frequency." This is not necessarily the exact middle of their frequency range. It is normally the frequency which makes use of the maximum energy emitted by the radar, since this represents the best potential for target detection (Fig. 2.1). Split frequency systems are slightly different. For split frequencies, known as "duo" or "dual detectors," see Chapter 13, Multichannel and Single Channel Systems. Whatever material is under investigation, it is important to make use of the maximum energy generated by the radar for detecting our targets. We therefore select this mid-frequency for calculation of the wavelength. In effect this means that we are calculating a wavelength at the peak energy of the antenna. We accept that there will be a range of frequencies on either side of the primary frequency and a range of wavelengths corresponding to these frequencies. However, to assess whether the selected antenna is capable

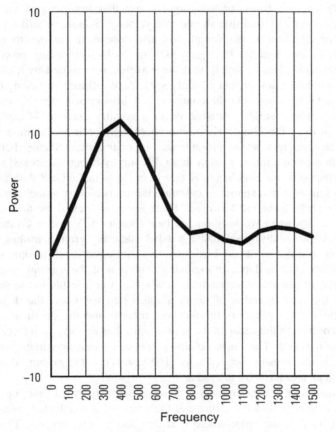

FIGURE 2.1 A frequency range for a 400 MHz GPR system. The profile shows that the quoted frequency occurs at peak energy.

of carrying out the tasks that we require, it makes sense to assess what is, in effect, the optimum operating wavelength.

VELOCITY AND WAVELENGTH

Radio waves do not travel at a constant speed. For this reason, radars measure depth in nanoseconds time, not meters and centimeters or feet and inches. It is a myth that measurement using a GPR is inaccurate. Radar measurement is extremely accurate. The inaccuracy, if there is one, lies in the translation of a depth in nanoseconds into a standard depth measurement. To make this translation we need to know the transmission velocity, the velocity at which the radio waves are traveling. This velocity depends upon the electromagnetic properties of the ground (or other material) through which the radio waves are passing (see also Chapter 5: Velocity Calibration).

It follows then from the equation $\lambda = v/f$ that wavelengths vary with transmission velocity and that should this velocity change, so will the wavelength. The variation in wavelength will also depend on the electromagnetic properties of the material through which the pulses are being propagated. This is not equivalent to saying that the wavelengths emitted by a radar are of unknown dimension or that a radar user cannot estimate wavelength when planning a GPR survey. On the contrary, it is important to be able to do so, particularly when working outside or in a country such as Malaysia, the Ukraine, or even the United Kingdom where heavy rain is common or in a wetland environment where groundwater is an important factor. Failure to make such a calculation can result in a GPR survey being unsuccessful.

The other obvious conclusion is that in order to make GPR data as accurate as possible, it is essential to calibrate the transmission velocity on every occasion on which the GPR is used. There are a number of methods of calibrating velocity when working on site (see Chapter 5: Velocity Calibration). Sometimes this is disguised under a label such as "ground truthing," i.e., excavating in order to obtain a comparative measured depth to one or more targets. Where this is done, an explicit calculation of the transmission speed may or may not be made, depending on whether it is possible to use software to adjust the data in terms of depth alone. Once we know the depth at a defined point in centimeters (or feet and inches) and the depth in nanoseconds according to the radar at the same point, it is an easy step to calculate transmission speed. The basic equation is speed = distance/time. Speed, in this case, is the transmission velocity. The distance is the probing depth and the time is as measured by the radar.

The radar's time measurement represents two-way travel time, i.e., transmission of the signal to the target and the return of the reflected signal from the target back to the radar where it is detected by the receiver. Therefore, the measured depth (d) from excavation equals one-half of the radar reading of time in nanoseconds (t) from the GPR survey. The velocity is therefore $2d/t$ in meters/nanosecond (or feet/nanosecond if working in imperial measures). Obviously, for this calculation to be accurate, it is essential that the calibration point must be the same for both the excavation and radar positions.

In reality, it is possible for the frequency to change slightly, depending on the ground conditions, since there has to be an interaction between the antennas and the ground. However, the variation is normally relatively small and can be ignored for the purposes of survey planning.

MAXIMUM DEPTH OF PENETRATION

As the signals pass into the ground, portions of it are either returned by reflection or lost when part of the energy dissipates into the ground (see also Chapter 7: Survey Strategies). In ideal conditions, the maximum depth of penetration before the full energy of the transmissions is lost is less than 20 wavelengths (Daniels, 2004). It is important to know this, if only to be able to recognize potential technical problems in advance and to know when a different antenna is needed for the task in hand.

In practice, there are few environments that have so little capacity for loss of the signal that a depth anywhere near the 20 wavelengths is going to be able to be achieved. One notable exception is ice. In most site conditions, but depending on the electromagnetic properties of the ground, the depth of penetration will be considerably less than 20 wavelengths.

Being able to calculate the wavelength allows the GPR operator to know whether or not the depth capability of the radar he/she is using is going to be sufficient for the task in hand. Although this is a general rule, it tends to be critical in a number of specific investigations, particularly:

- checking on targets below reinforced concrete;
- locating targets in wet or waterlogged ground;
- any very deep investigation.

One obvious implication of knowing the limitation on the amount of energy emitted by a radar is that deep investigations will require longer wavelengths in order to reach the probing depth required. Wavelength is inversely proportional to frequency so the longer the wavelength, the lower the frequency of antenna. It follows then that deep investigations require lower frequency systems.

THE SMALLEST TARGET WHICH CAN BE DETECTED

No single radar can detect targets regardless of their size. The size of the wavelength determines how small a target can be detected. It is not true, as has sometimes been claimed, that the size of the target does not matter. Since radars can only measure in fractions of their wavelength, an object which is smaller than that minimum size will simply remain undetected (and undetectable). To be detectable, a target should be at least approximately 10% of the wavelength or greater than this. It will also have to be physically distinguishable from other surrounding targets (see Target Spacing—How Close/How Far Apart section).

The implication for all targets is that a short wavelength is preferable for finding and/or defining small targets. However, this automatically provides a potential conflict with the depth requirement outlined above since a short wavelength will limit the depth to which probing can be carried out. There is therefore often a conflict between the radar requirements for the size of targets to be detected and the depth at which these targets lie.

Provided that a small target is close to the surface of the material under investigation, it can be detected by a high-frequency GPR. A small target that lies deep underground cannot be detected with a high-frequency GPR since the electromagnetic pulses cannot reach the full depth. It will also remain undetected by a low-frequency GPR if the target size is below 10% of the wavelength, even though that radar might be capable of achieving the depth required.

The exact parameters will depend not only on the type of target but also on the electromagnetic properties of the soil (or building material) in which they lie. This is why it is important to be able to at least estimate the wavelength of the GPR which is going to be used in the investigation. An experienced operator will realize that he/she needs to be reasonably sure that the target size and depth are compatible with the radar that he/she intends to use for the survey. The few minutes spent in calculating possible wavelengths is never going to be time lost since this may alert the operator to the need for a different frequency of antenna.

TARGET SPACING—HOW CLOSE/HOW FAR APART?

The space between targets is as important as the size of the target when using a GPR. It is possible either for one target to obscure another or for other buried material to hide the existence of a target. As mentioned earlier, the radar measures in function of its wavelength. If one target lies too close to another, the signal responses of the two targets merge. From the perspective of the radar, they simply become one target. Targets within a half wavelength of each other will result in a combined signal (Daniels, 2004). On screen the radar user will see one single hyperbola, not two or more. A good rule of thumb therefore for both targets to be detectable is that they should be separated from one another by 1 wavelength. It follows that, for closely spaced targets, a higher frequency (and shorter wavelength) radar will be required.

Fig. 2.2 shows the layout of a number of buried utilities in the ground. This is not untypical of the subsurface in a major city. Not only do the utilities cross each other, some of them lie immediately adjacent to one another. The radar will be able to distinguish those that lie at different depths. However, unless a very high frequency of antenna is used, it may not be possible to detect those that lie close to one another. In the simplest case, two or more cables which lie close to one another will show as one single (merged) hyperbola.

The typical type of practical result that this would give in radar terms is shown in Fig. 2.3. This shows two consecutive profiles of GPR data, collected along parallel survey lines at 500 mm distance from each other, across a pavement below which lie a group of closely spaced lighting cables. It is clear from the data that there are multiple utilities in the immediate vicinity of the line indicated by the marker "1" but it is not at all clear how many there are or the precise depth of each one. The variation in patterning from the first to the second radar trace is as unexpected as it is confusing since it is unlikely that the depth and number of cables actually varies by any great

FIGURE 2.2   Excavated utilities.

amount. The changes in patterning are caused primarily by all of the cables lying closely together, so that any slight variations in the distance between adjacent lighting cables result in each hyperbola in the data representing one or more of the cables but not necessarily exactly the same ones as the previous radar trace. The gaps between the lighting cables are simply not large enough for the wavelengths emitted by the radar used. The resulting images are very difficult to interpret beyond saying that there are a number of cables in the vicinity of the marked position.

This data could be resolved by using a higher frequency antenna, one whose wavelength was shorter than the average gap between the lighting cables. Note that this is not a case of GPR failing to detect but of the wrong frequency of antenna being applied to the problem. There will, of course, be instances where it is not possible to apply the appropriate antenna frequency, notably where the utilities lie outside the depth range of the most suitable antenna.

This is also an extremely important principle for anyone wishing to examine the subsurface of reinforced concrete. Radio waves cannot penetrate

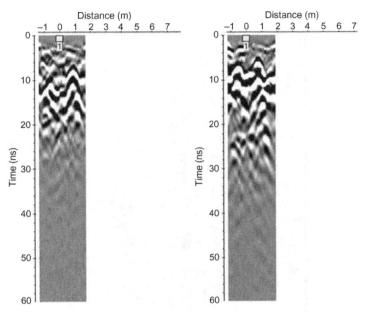

FIGURE 2.3   Two parallel lines of radar data taken across closely spaced lighting cables.

metal. If metallic grids have been used as reinforcement, then it is important that each rebar is spaced at no less than 1 wavelength from its neighbor. If the spacing is less than a wavelength, then the signals will merge and the reinforcement layer will appear to be continuous, as though there were a metal sheet in place.

Fig. 2.4 shows an area where the surface at the start of the survey is reinforced concrete which then gives way, after a distance of approximately 12 m, to asphalt (marked as "tarmac" in the image). It is easy to see in the area below the asphalt that there are real signals from anomalous material buried below the surface because the signals lower down do not in any way resemble those above. The data are a real reflection of the objects in the subsurface. In the concrete area, however, it is equally clear that some sort of echo effect is occurring. This is because the spacing between the reinforcement bars in the concrete is smaller than the wavelengths emitted by the radar. The result is an echo effect, known as ringing, from what the radar perceives as a metal sheet. As the radio waves cannot pass through the reinforcement in the concrete into what lies below, there is no subsurface information from this area and the radar output is characterized by echo effects known as "ringing" instead.

Contrast the situation where a very high-frequency radar (4 GHz) has been used on reinforced concrete (Fig. 2.5). Here we can see very clear definition of the reinforcement bars (rebars), the reinforcement mesh below

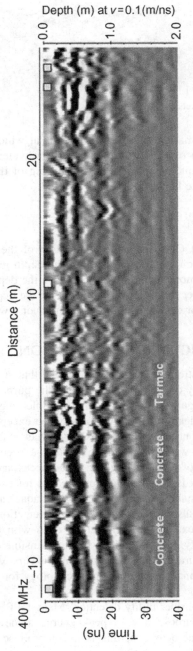

FIGURE 2.4 Ringing (echo effects) from surveying rebars with a 400 MHz antenna.

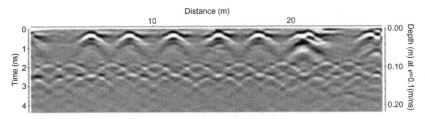

FIGURE 2.5   4 GHz definition of rebars.

(shown as hatching in the image), and even the points at which two of the rebars lie very close to each other. This is because the wavelengths emitted by the 4 GHz antenna are much smaller than the spacing of the rebars and are therefore able to penetrate between the bars.

DEPTH ACCURACY

It follows from the fact that GPRs measure in function of the wavelengths they emit that there is an accuracy associated with the depth measured. The shorter the wavelength, the more accurately it will place the depth of the target. In other words, the higher the frequency, the greater the depth precision. The general rule of thumb for depth accuracy is a quarter of a wavelength.

FITTING THE FREQUENCY TO THE APPLICATION

It follows from the observations made in this chapter that it is extremely important to fit the frequency of the antennas used to the purpose and to the environment of each investigation. Although it is possible to generalize for the majority of specific applications, there will always be exceptions and the user is faced with the risk of nondetection of his/her targets or finding out by painful experience that the work needs to be tackled a second time and sourcing another frequency of antenna is going to be necessary. Given that safety in one form or another is the most common reason for using a GPR, it is important to be able to establish whether or not the radar can reasonably be used for the survey and also what the risks of nondetection are. As we will see from the practical examples given below, even with the unknown factor of transmission velocity, it is possible to make a sensible evaluation of what GPR equipment is required, rather than going into a survey blindly with the wrong frequency of antenna and blaming the technology for the failure to detect all targets.

As an example of this, consider utility detection. Many GPR systems marketed specifically for this purpose have a peak energy somewhere around 800 MHz. As for much radar detection, this is a compromise between the depths at which most utilities lie and the size of targets whose detection is needed. Although this is an extremely useful frequency for the majority of pipes and services, a change of frequency will almost certainly be required if it is necessary to detect either a deeply placed service or one which is lying below open ground known to be wet at the time of the survey. Typically both of these conditions can apply in the detection of foul water sewage pipes.

At the other end of the scale, near surface cables may require the use of a higher frequency antenna. Although detection of plastic pipes is frequently a matter of concern to the utility companies, there are typically more strikes on near surface cables. This implies that small targets close to the surface are not always detected which, in turn, suggests that the use of a high-frequency antenna (e.g., 1.5 GHz or above) as well as the traditional lower frequencies would improve the protection of fiber-optic cables already in the ground.

Another example is the prevalence of 400, 500, and 250 MHz antennas in use in archeological investigations, the latter typically intended to allow for the moisture content of open ground. The full range of archeological surveys covers many different environments and archeological investigations can include the interior of buildings, historically important churches being one obvious example. It would not be appropriate to take a 250 MHz GPR into the interior of a small church chancel. Not only would it be unlikely to resolve any near surface targets with any accuracy, there would be a risk that the beam spread would cover the dimensions of the floor before any great depth had been reached, leading to apparent anomalous material which, in reality, represented reflections from the outer walls.

It is important to plan the survey in function of the information needed and the site conditions. Using a specific radar because it happens to be the one available is not a good strategy.

PRACTICAL EXAMPLES OF WAVELENGTH CALCULATION

Detecting Through Reinforcement

A client requests a GPR survey of an area of reinforced concrete to establish whether there are any utilities present in the subsurface. You have a single channel GPR system with a choice of 1 GHz, 500 MHz, and 250 MHz antennas. How would you choose between the antennas?

The critical factor here is the spacing between the rebars in the concrete. There is no point in even considering maximum depth or whether the frequency matches expected target sizes and depths until you have made sure that the signals can pass through the gaps in the reinforcement. The result of carrying out the survey using a radar with too long a wavelength will give data similar to Fig. 2.4. The first step is to establish, if possible, the rebar spacing. Assume that the information from site gives a rebar spacing of 125 mm.

Although we do not have a transmission velocity for this site, it is not uncommon for the ground directly beneath man-made surfaces such as concrete and tarmac to be dry. This means that, for planning purposes, it is reasonable to assume a velocity of 0.1 m/ns.

Using the formula $\lambda = v/f$, where λ is the wavelength in meters, v is the transmission velocity of the radio waves transmitted by the radar, and f is the frequency of the radar in GHz, we can calculate the wavelengths (at peak energy) for the three sets of antennas we have available. The reader is advised to do this for him/herself before continuing.

For the 1 GHz antenna, $\lambda = 0.1/1$ or 100 mm. For the 500 MHz antenna, $\lambda = 0.1/0.5$ or 200 mm and for the 250 MHz antenna, $\lambda = 0.1/.25$ or 400 mm. It is therefore obvious that the only antenna that we can use from this group is the 1 GHz since it is the only one to have a wavelength smaller than the spacing of the reinforcement bars. This is important not only for being able to detect below the level of the reinforcement but also because the shorter wavelength places a depth limitation on the detection capability of the radar. It would be extremely foolish to commit to detecting a target below 2 m, for example. Why 2 m? The reader is advised to calculate maximum potential probing depth, based on the information given earlier in this chapter.

Even if the rebar spacing is unknown, this problem can be tackled by using a high-frequency antenna to determine the spacing of rebars on site.

If, instead of calculating the wavelengths, we had decided that a 500 MHz antenna would provide sufficient depth and good target definition, we would have been sorely disappointed and left with a survey in which it appeared that GPR had failed. The reality is that we, as the equipment operators, would have failed to notice the critical element in the information we had been given. Equally reaching for any radar to do this survey without calculating the wavelength would not have been a sensible option. In GPR surveys, wavelengths always matter.

Detecting in Wet Ground

The presence of water makes a significant difference to the transmission velocity and, because of this, the wavelengths emitted. It also therefore affects the depth to which the radio waves can effectively penetrate. A client has asked us, as part of an archeological investigation, to find an underground vault. In the vicinity of this feature, there should be an early Victorian lead water pipe. As with most archeological sites, the remains are not thought to lie any deeper than 2 m at most. Although there was a building on the site, this has been in ruins for years and the majority of the site is therefore uncovered. It is currently very wet under foot.

The first problem is that, not having been on site, we have no means of knowing the transmission velocity and it would clearly not be appropriate to assume 0.1 m/ns. In this case, it is more likely that we will know which frequency we intend to use and 250 MHz, as outlined above, is a likely candidate. If we assume a velocity of 0.07 m/ns, this will give us a wavelength of 0.07/0.25 or 28 cm. The comparative calculation for a dry soil (for which the transmission velocity would be c. 0.1 m/ns) would be 0.1/0.25 or 40 cm. So the depth penetration of the radar could potentially have been reduced by 30%. It is unlikely that we would be able to reach the full 20 wavelengths due to ground losses but, in any case, the water content has significantly reduced the depth to which we can probe.

If we repeat the calculation, using 0.05 m/ns, we can see that the wavelength reduces to 20 cm. So, even without knowing the transmission velocity, calculating possible wavelengths provides a warning that using a lower frequency radar is not necessarily sufficient, we have to consider the depth setting also. This is something we will look at in more detail in Chapter 4, The Effect of Water and Air.

REFERENCE

Daniels, D.J., 2004. Ground Penetrating Radar, second ed. (Chapter 2), pp. 22, 23, 30.

Chapter 3

More Fundamentals of GPR Operation

Chapter Outline

GPR DETECTION

GPR is a powerful technique because, within certain limits, it recognizes the changes between materials. It is not dependent on the presence of a particular type of material. Instead it reacts to the changes in electromagnetic properties between any materials. Provided there is a change either in electrical properties or in magnetic responses, any target is potentially able to be detected.

Unfortunately, as human beings, we use our eyes, optical detection, and are therefore used to distinguishing materials on a totally different basis. It is important to understand that GPR cannot distinguish any of the following characteristics without additional information:

- The nature of the target material
- Color
- Density
- Small objects obscured either by larger ones or by the proximity of similar material
- Which type of utility is represented
- The date of archeological remains.

It is possible to infer some of this information by ancillary means, normally above ground markers or externally provided information. For example, the radar operator may be standing on an asphalt surface but be able to see on screen a series of regularly spaced hyperbolas within the next layer down from the surface. This patterning suggests the existence of regularly

Ground Penetrating Radar. DOI: http://dx.doi.org/10.1016/B978-0-08-102216-0.00003-5

spaced reinforcement bars from which we could infer that there is a layer of reinforced concrete below the asphalt (cf Fig. 2.5). In utility detection, tracing the line of a pipe, cable, or duct to an above ground marker, an electricity substation or a known water supply or drainage point will enable the nature of these utilities to be identified. It may be possible to identify either air gaps or metal on the basis of signal strength (amplitude) and/or echo effects. Patterning in the spatial layout of the GPR data may give clues to the existence of particular types of feature but the radar data itself does not give this information. In general terms all radar signals are the product of change and one of the complications is that the same level of signal response may be the result of different combinations of materials.

The concept of density is a particular problem since it figures from time to time in GPR reports. Since density is not a feature of electromagnetic response, no radar can measure this. It is inappropriate to describe GPR targets as being more or less dense. Working out exactly what the radar is responding to within the context of a particular site is one of the challenges of interpreting GPR data. Usually obtaining a realistic understanding of the GPR's response relies on examining the patterning in the radar data, both as seen in the two-dimensional (2D) vertical view and, where possible, horizontally, extracted from a 3D data set. It also means not only looking at the anomalous signals themselves but also their continuity from one data set to the next. If a case is to be made for signals of the same strength (signal amplitude) as representing the same materials, there needs to be evidence to support this.

WHAT 2D GPR DATA LOOKS LIKE AND WHY

A typical single GPR signal passing into the ground looks something like Fig. 3.1. The precise shape and duration of the signal will depend on the design of the radar used, the operating frequency, and also the ground conditions. The signal generated by the GPR begins on the left-hand side of the image and continues down into the ground toward the right.

The signal is composed of positive and negative parts, depending on whether the precise point selected lies above or below the zero axis (shown by a red (black in print versions) dotted line). These are the parts of the signal that give rise to the black and white banding, black representing the positive parts of the signal and white the negative. Single signals are sometimes referred to as "A-scans" or "wiggle traces."

The direct portion of the signal represents the direct signal from transmitter (Tx) to receiver (Rx) and is commonly taken as marking the position of the ground surface. Before this (i.e., to the left in Fig. 3.1) there is a small amount of signal which essentially marks the transition from the generation point in the controller down into the ground, via the antenna. This is known as the time zero, usually abbreviated to Tz. Its importance lies in its removal since it forms no part of the depth below the surface. If it is not removed

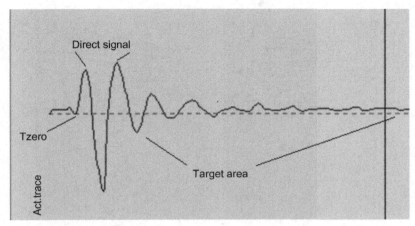

FIGURE 3.1   A typical GPR signal.

from the data, it will exaggerate the depth of any targets. Tz is measured at the first zero crossing along the axis before the direct signal.

Below the direct signal lies the full remaining depth of probing which forms the target area. It will be seen later that it is useful to be able to identify this, if only for optimizing data processing (Chapter 8: Data Processing). As the radar moves along the ground, signals are emitted by the transmitter at a predetermined sampling rate. The shape of each signal returned to the receiver is determined by the materials in the subsurface, including any targets. The typical 2D data profile or radargram is built up from these individual traces (Fig. 3.2). In this example, Tz has already been removed. Had it not been, there would be a layer of uniform gray at the top of the gray scale plot, the uniformity forming a clue to the fact that it does not represent subsurface data.

This 2D image gives an outline of what lies in the subsurface. Although the figure shows this data in gray scale, it can be depicted in color. If color is used, it is useful to be aware of the potential limitations (see below). Radargram is becoming a commonly used term currently for this 2D vertical depiction of GPR data but the older term of B-scan is sometimes also used as are the simple terms 2D data, or radar profile, or radar trace. The formation of 3D data from 2D data is dealt with in Chapter 11, Common GPR Applications.

WAVELENGTH AND SIGNAL AMPLITUDE

Fig. 3.1 also illustrates the concepts of a wavelength and signal amplitude. Signal strength or amplitude is measured by the distance of each curve from the zero axis (or red (black in print versions) line in the figure). The direct signal is very much stronger than those of the target area. The signal amplitude of the first part (to the left) of the target area is correspondingly larger

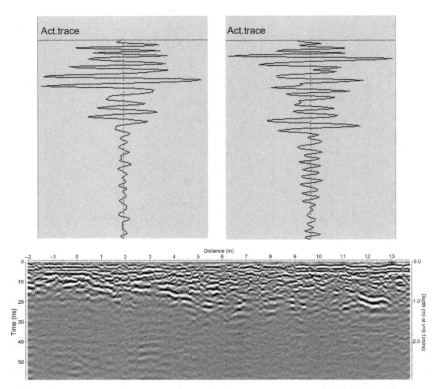

FIGURE 3.2   The gray scale image below is built up from the changing individual signals such as the two individual signals shown above.

than that of the lower levels (to the right). This is to be expected since as the signals reach each change in materials, a portion of the signal is returned to the receiver, leaving less to continue into the ground.

The wavelength of the signal is measured from one zero crossing to the second subsequent zero crossing, i.e., it includes one positive and one negative section of the wiggle trace. Starting at the point marked "Tzero" in Fig. 3.1, the wavelength runs to the point at which the signal sweeping upward into the positive section crosses the zero line. The signal amplitude varies with depth and with the nature of the buried material. The wavelength will often be a constant but it varies if the transmission speed changes.

NONHYPERBOLIC TARGETS

We have already seen in Chapter 1, Introduction to GPR, how a single target, viewed in 2D GPR data, forms a hyperbolic shape. The cone shape of the radar beam detects subsurface material before and after the radar arrives directly above the target. The distance from the radar to the target therefore decreases as the radar approaches, reduces to a minimum when the radar is directly above and then increases as the radar moves on. This forms a hyperbolic shape in 2D data.

It is very important to understand that the hyperbola is formed by the shape of the radar beam and the transmission velocity of the electromagnetic pulses, not the actual shape of the buried features. The exception to this is when the feature is much larger than the wavelength. Under these circumstances the outline, made up from a number of hyperbolas, will follow that of the feature. In essence, the radar data will be composed of a number of readings which follow the outline of the target. The shape of the individual hyperbolas is still determined by the radar design and the transmission velocity. The beginning and end of the outline will reveal the end hyperbolas. The same process is observable in the outline of stratigraphic layers: see, for example, Fig. 3.2 where some layering is also visible.

It is quite common, however, for targets not to appear as hyperbolas but as irregular or squared off shapes. This can often make them more difficult to interpret, particularly when surrounded by other buried material. To consider why this might happen, look at the archeological example in Fig. 3.3.

The image shows a line of GPR data collected across the foundations of three double stone faced walls with robber trenches on either side. Robber trenches are formed when the stone from an abandoned site is excavated for reuse. The interpretation of stone walls and robber trenches comes from the excavation of the site, after the GPR survey had been completed. Two of the walls lie close together (on the left-hand side) and the third one further away. Between the second and third walls there are lots of small signals, indicating archeological material left in situ. Some of this is represented by traditional hyperbolas but other signals are more irregularly shaped. The walls themselves are characterized by columns of hyperbolas up to a certain level. Above this level the column continues but the hyperbolic shape has gone, replaced by short blocks of stronger signal amplitude, indicated by a greater degree of black and white contrast. The depth level at which this change occurs is the point at which the robber trenches meet the wall foundations. This is not a coincidence.

Robber trenches are typical of the exploitation of a historical site after it has fallen into disuse, especially where building materials were expensive or not

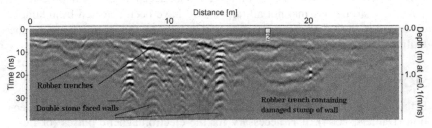

FIGURE 3.3 2D GPR data across the footings of a Romano-British villa.

readily available. The trenches were dug in order to remove and reuse the stone from abandoned buildings. Above the line of the robber trenches, the remaining stone has been damaged as other stones in the wall were prized out and removed. Below this same line, the wall footings remain undisturbed and intact. The smooth, uninterrupted hyperbolas indicate exactly that: an undisturbed, intact, smooth surfaced target with no similar material in close proximity. These smooth hyperbolas are also a good example of how GPR data represents any target as a hyperbola rather than depicting the outline of the material in the way that a camera would. It is also worth noting that the double stone facing is not detected by this frequency of antenna which simply "sees" the two stones as one entity. Either there is no gap between the two sides of the wall or any gap is too small (relative to the wavelength used) to be detected.

Above the line of the trenches, the distortion in the signal is the result of multiple reflections from the damaged, irregular surface, causing localized interference in the signals returned to the receiver. This interference enhances the signal in certain places, resulting in the stronger contrasting blocks of black and white. In others, it cancels out part of the signal, resulting in the loss of definition of parts of the hyperbola.

The same effect can be seen anywhere where either damage has occurred or there is a cluster of material adhering to the target or surrounding it. For example, the material surveyed in Fig. 3.2 comes from a different site but one which contains very similar material. Given that even modern features buried beneath the ground surface, such as pipes and cables, will typically have other material in their immediate vicinity, it is not uncommon to see this type of truncated signal rather than neat hyperbolas in 2D radar data.

This has important implications for many GPR investigations. Depending on ground conditions, it may or may not be possible to detect targets simply on a single pass of the GPR. Fig. 3.2 shows data collected on open ground but data collected over man-made surfaces such as asphalt can appear similarly complicated. If it is not possible to pick out all of our targets because they are not necessarily neatly delineated, then we have to look to other measures to be sure that we have located everything that we set out to detect. It is for this reason that many experienced GPR surveyors do not consider that marking up the position of buried targets at ground level, on the basis of what the GPR surveyor can see on screen is an acceptable survey procedure. There is a strong risk that not all targets will have been detected from simple observation during the survey. We will return to this topic again in the Utility Detection section of Chapter 11, Common GPR Applications.

SIGNAL CAPTURE AT THE GROUND SURFACE

Effective transmission into the ground is essential for efficient target detection (Daniels, 2004). It is necessary for the electromagnetic pulses (or radio waves) to be captured as fully as possible at the ground surface with as low a level of reflection as possible. In simple terms this means that the antenna or antennas of most GPRs should always lie flat to the ground surface, in which case they can be said to be ground coupled. This rule depends on the antenna design and assumes that the antennas are themselves a flat shape. This represents the most common form of GPR antenna.

In this context coupling simply means that the antenna is optimized for transmission into a certain type of material. For GPR this is most commonly the ground but antennas can be electromagnetically matched to air or water also. If they are used in contact with the air then they are said to be air-coupled, or water coupled in the case of use directly in contact with water, and so on for any other type of material. The electromagnetic matching is an important design feature of any radar. It is possible to set up some unusual and unhelpful effects which will mask the data the operator requires if the antennas are not properly matched to the surface material under investigation.

An important early study of GPRs, carried out by the Building Research Establishment in the United Kingdom, tested the degree of reflection and absorption of the transmitted signals and found that antennas which were not ground coupled lost over 50% of the signal in above ground reflection (Matthews, 1998). This finding is independent of the radar design.

In reality, ground surfaces, even relatively flat man-made ones (e.g., asphalt, concrete, or laid stone floors), are rarely completely flat. There is also the question of using GPR over open ground where the surface may rise, fall, or undulate and of certain GPR applications where traveling at speed may be necessary.

The first rule of thumb with a rough surface is to travel slowly enough for the antenna not to lift off the surface. If the antenna lifts, there is a risk of creating an artificial artifact from the air gap between the antenna and the ground. Air gaps are often the source of echo effects. Should the air gaps cause echo effects, these secondary signals risk being misidentified as subsurface features. Where the signals are the direct result of subsurface features, there is also uncertainty of where these are located since the angle at which the radio waves are being transmitted is no longer directly into the ground.

One other possible solution is to use a lower frequency of antenna than one would normally. The longer the wavelength, the better the signal capture at the ground surface. The size of the antenna is directly related to the size of the wavelength(s) transmitted and so a lower frequency antenna will be larger and heavier than its higher frequency equivalents. The extra weight implied by its larger size generally results in a greater ability to remain in close contact with the ground along the line of travel. Target definition will suffer as a result of using a lower frequency antenna: all targets will be smaller relative to this longer wavelength, including any artificially created. There is also an increased risk of not detecting small targets.

Where high-speed operation is required, it may be necessary to raise the antennas above the level of the ground. This tactic is frequently used in road inspections and also in railway track surveys where it is necessary for the survey vehicle to travel at speed. If the antennas were to remain in contact with the ground, they would very quickly become damaged either from detritus left on the road or track surface; from variations in the road or track surface; or from external features built into the roads and tracks (road or track "furniture"). Although the proportion of signal reflected at the surface inevitably increases and the amount of signal available for detection is lessened, it is still possible to operate the GPR effectively, provided that the antennas are placed closer to the ground than a quarter wavelength. In practice, it is possible to set up resonances if the antennas are a quarter of a wavelength from the ground. Experience suggests that halving this distance from the ground surface works well. Where it is necessary to raise the antennas, placing the antennas at or within one-eighth of their wavelength from the ground surface is an effective strategy. Another option is to use an air-coupled antenna. This ensures good transmission into the air but does not necessarily reduce reflection at the road surface.

The use of GPR on water is another example of where care should be taken over coupling. GPR works well in freshwater. It is common practice to place a ground coupled radar inside a boat and survey through the bottom of the boat. This is often successful even though the electronic performance of the radar is effectively matched to ground conditions, rather than water. However it is also possible to set up interference effects from transmitting through two surfaces without being electromagnetically matched to either. One option to prevent this happening is to use a water coupled GPR directly placed floating in the water.

SIGNAL-TO-NOISE RATIOS

The objective of any GPR survey is to obtain as clear an image of the potential targets as possible. All 2D data contains a certain level of "noise," i.e., unwanted signals whether this is background ringing, signals introduced from carrying out a survey over a rough surface, or echo effects from subsurface material. The signal-to-noise ratio is a way of describing the clarity of the data and is, in part, a feature of the design of the GPR system. High signal-to-noise ratios have clear data for which analysis will be much easier than data with low signal-to-noise ratios. Low signal-to-noise ratios are a risk of surveying from very uneven surfaces.

DATA PRESENTATION: COLOR AND GRAY SCALE

Although it is still common place to display GPR data in gray scale, displaying the data in color is becoming more popular. Since people are naturally oriented toward optical detection, potentially this is more comfortable to look at than swathes of gray scale data. Color can also sometimes be used to highlight areas of interest more clearly than the monochrome version. However, some care has to be taken to make sure that (1) the data is not overinterpreted, (2) the interpretation makes sense in terms of electromagnetic pulses rather than in optical terms, and (3) that some of the fainter targets are not lost.

The point at issue here is the difference in depiction of the relative strength of the returned signals. Signal strength is measured by the amplitude of the signal and is the measure of difference between any material and those around it. Gray scale is normally a graduated scale either from black (positive) to white (negative) or from black to black where all stronger signals are denoted in darker shades. There is a wide range of color scales available, two examples are visible alongside gray scales in Fig. 3.4. Where there is a continuous level of difference between a feature and its environment, there is no difficulty in

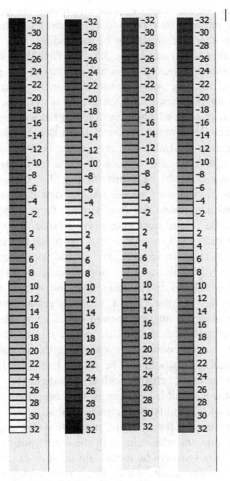

FIGURE 3.4   Four examples of signal amplitude scales.

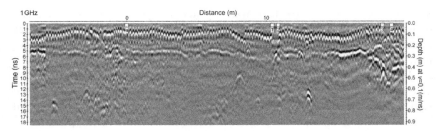

FIGURE 3.5 GPR data illustrating changing signal strength (amplitude) along the base of reinforced concrete.

using any of these scales. Where gray scale can be an advantage is where the feature is continuous but the value of signal amplitude changes because of the intrusion of another material into the vicinity. In this case, the sudden change from one color to another can sometimes be misleading.

An example of this is shown in Fig. 3.5. The first layer of strong signals comes from the multiple hyperbolas toward the top of the plot. This rather random pattern indicates very poorly reinforced concrete. Instead of the reinforcement grids being laid across the surface in a systematic fashion, they appear to have been positioned haphazardly, in places being forced one on top of another. Directly below the concrete, there is a continuous line across the radargram indicating the interface between the concrete and the material below. For this data, a gray scale of black to white (as in the first column of Fig. 3.4) has been used. At the beginning of the plot, the signal strength of this lower interface (lhs) is very strong for the first c. 2 m. Thereafter the contrast in the signal designated this line is much fainter, indicating reduced signal strength and the line almost disappears from c. 8 to 10 m along the line of travel of the radar (left to right across the plot). For the last c. 5 m of the line, the interface is again very clearly visible with strongly contrasting bands of black and white. This variation in gray scale intensity indicates that signal strength varies significantly along this line and is strongest at the beginning (lhs) and toward the end of the radargram. From this we can deduce that another material may be present in the areas where the signal is strongest, almost certainly water. The use of gray scale allows the person interpreting the data to see continuity even where the signal amplitude varies. Although the continuity would still be visible had color been used, it can be a distraction that the colors change with amplitude as they would do along this line. Where color is used, it depends on the abruptness of the change as to whether this type of continuity is as clearly visible.

Color can be very useful, for example, in emphasizing discrete strong signals but there is an inherent arbitrariness in the cutoff points for the individual colors which may not always be useful. As another example, think of the wearing course of a bridge across which lies a road with an asphalt surface.

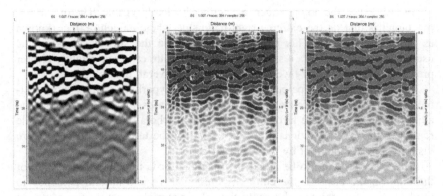

FIGURE 3.6 The same data presented in gray scale and two color variations.

It will be possible to use a GPR to identify the base of the asphalt layers and, with a sufficiently high-frequency antenna, the base of the wearing course. Assuming that the wearing course is in good repair, the signal returned from this level is likely to be quite weak because there is little change in materials from the wearing course to the asphalt beneath. However, where voids have built up, usually from rain seeping in through cracks in the surface, the signal will be very much stronger. This is because, in place of a simple asphalt-to-asphalt interface, either air or water may be present as well. The relative permittivity of air is approximately 1 and of water approximately 80. The presence of either will change the amplitude of the signal by a large amount. It may be helpful to have this highlighted in a bright color, especially if the survey is being carried out to check on the development of voids. However, if the whole wearing course layer is to be mapped, then this may be less helpful since the asphalt-to-asphalt interfaces are unlikely to be as easily distinguished. Some software processing packages allow the color changes to be set according to the survey site so as to minimize the possibility of this being a problem.

This problem may also occur in other contexts such as relatively faint signals reflected from a buried target. Fig. 3.6 shows how easy it is to lose faint targets and also how this may vary with the color scheme used. These data come from a GPR survey to locate buried utilities and the principal target is a plastic pipe lying below an asphalt road. In the gray scale version, there are two faint hyperbolas visible toward the right-hand lower corner of the data, the upper one being indicated by a red (gray in print versions) arrow. The reason that these two utilities are so faint is because the upper layers of tarmac are strongly responsive and a large part of the transmitted signal is being returned from these layers. Even so, these two hyperbolas are reasonably clear to the eye.

In the first of the color sequences, neither is clearly visible. The only obvious target is that on the extreme right-hand side where an echo effect (known as ringing) is visible. This is unfortunate because the simple color scheme used to present these data results is a much more attractive and slightly clearer presentation of the upper layers and hence of the overall data plot.

Increasing the color palette makes the upper hyperbola slightly more visible in the second colored data set. It also increases the visibility of some adjacent material (to the left) but the lower hyperbola on the right-hand side remains invisible. If color is used, care needs to be taken that all targets are visible and not faded out by the resulting contrast.

The other problem to be avoided if using color is the misinterpretation of the data on the basis of the color responses. There is a natural tendency to think that, if the color is the same then the materials must be the same. In radar terms, this translates to "if the signal amplitude is the same, then the materials must be the same". This is not the case. The same signal amplitude (or the color which indicates it) can be generated by completely different sets of materials. The radar does not measure the response from a single type of material. It measures the change from one to another.

There is no right or wrong in using color or gray scale. One useful compromise is to use both gray scale and color for different stages of the processing. For example, gray scale may be used for the purpose of analyzing the data then a color scheme introduced to present the results for the benefit of the client, making sure in the process that significant results are not omitted due to an abrupt change of color or the lack of a color change. In this way the gray scale data can act as a check on the color presentation.

REFERENCES

Daniels, D.J., 2004. Ground Penetrating Radar, second ed. (Chapter 1), p. 5.

Matthews, S.L., 1998. Subsurface radar as an investigative technique, Building Research Establishment Report BR 340, CRC, Garston, Watford.

Chapter 4

The Effect of Water and Air

Chapter Outline

It is sometimes suggested that radar data are "inaccurate." This is not the case. Radars measure extremely accurately in nanoseconds time. However, for these depths to be meaningful to anyone using the data, they require to be turned into meters and centimeters or feet and inches. This will not always be simple to do because it is possible for the subsurface to contain mixtures of materials sufficiently different in electromagnetic terms for the velocity to change in value either with depth or across the survey site. Where GPR data prove to be inaccurate in terms of depth, it is usually the case that the conversion from measurement in time to absolute depth was the cause rather than the original measurement being inaccurate.

TRANSMISSION VELOCITY AND MISLEADING DEPTH MEASUREMENTS

Radio waves do not travel at a constant speed. Their transmission velocity depends on the type of material through which the electromagnetic pulses are traveling. In particular, they depend on the electrical and magnetic properties of this environment. The velocity will therefore be different for asphalt, concrete, open soil, stone, brick, air, water, or any other material in which the targets lie buried.

The situation becomes slightly more complex when there are mixed materials present. The two materials which will distort the transmission velocity more than any other are water and air. This is because their electromagnetic responses are very different from those of dry soil or building materials. The relative permittivity (or dielectric constant, Er) is a measure

Ground Penetrating Radar. DOI: http://dx.doi.org/10.1016/B978-0-08-102216-0.00004-7

TABLE 4.1 Comparison of the Relative Permittivity of Dry Soil With Those of Water and Air

Material	Dry Soil	Water	Air
Relative permittivity (approx.)	9	81	1

of the electromagnetic response of any given substance. Table 4.1 provides the difference in Er for dry soil, water, and air. Although both air and water distort the transmission velocity, it is evident from the high value of Er that water provides the greater potential for distortion as well as being the material most likely to be present in any quantity, at least in underground contexts.

The relative permittivity of the material being surveyed determines the transmission velocity of the radio waves through the survey environment. The relationship between this velocity and the electromagnetic properties can be stated as:

$$v = c \div \sqrt{Er}$$

where v is the transmission velocity, c is the speed of light, and Er is the relative permittivity of the environment through which the radio waves are being propagated.

This simple equation is useful because it allows us to calculate potential transmission velocities for different sites. If we know or can approximate the transmission velocity, that allows us to estimate potential survey depths in radar terms in advance of survey. It also allows us to adjust those calculations to take account of changing ground conditions or to test whether, for example, there might be an air gap present in our data. For these purposes, it is not necessary to use an exact value for the speed of light and the value of c can be approximated to 0.3 m/ns. The equation can therefore be restated as:

$$v = 0.3 / \sqrt{Er}$$

THE "RULE OF 3"

Before going on to consider how we can apply this formula in practice, there is a very simple but useful rule of thumb, the rule of 3. The relative permittivity of air is 1. In air, radio waves travel at the speed of light which, for simplicity's sake, we can approximate to 0.3 m/ns. Dry soil has a dielectric permittivity of approximately 9 which then implies a transmission velocity of 0.1 m/ns. Freshwater has a relative permittivity of 80−81 which gives a transmission velocity of 0.033 m/ns. This gives a useful approximation rule of 3 which can be helpful when planning the probing depth of the radar.

Propagation through air is three times faster than propagation through dry soil.

Propagation through dry soil is three times faster than propagation through water.

WET SITES

Given a situation where a survey is to be carried out over open ground which is known to be wet, then the worst-case scenario is that three times the normal probing depth will be required, the normal probing depth being the depth required in dry soil. If we carry out this survey without considering the effect of groundwater and use the same probing depth as we would have done in dry soil, it is extremely unlikely that we will find the targets. There is a good chance that the targets will lie below the probing depth. In this scenario, it is not unusual for the operators to then claim that it is the technique of GPR that is at fault and that the targets could not be found because "GPR does not work in water." This is an invalid claim as GPR works very well in freshwater. The difficulty lies with the team who did not consider whether their equipment met the probing depth requirement for the task.

As an example, consider the case where the survey target is a foul water sewer which passes below a field. Either from the information given before the survey or from surveying over adjacent dry ground (e.g., below concrete or asphalt), we believe that the sewer may be buried to a depth of 2.5−3.5 m, the variation in depth being due to the sewer passing below at least one drainage ditch. The radar to be used is intended for utility detection and has a maximum range of 90 ns, essentially 4.5 m in dry soil. To see how this is derived:

Speed = distance/time
The distance traveled by the signal = 2 × depth of target
(The signal must be transmitted to the target and then returned to the receiver antenna.)
Speed = 2 × depth/time
2 × depth = speed × time
2 × depth = 0.1 × 90 = 9 m
Maximum target depth = 4.5 m.

The maximum target depth is therefore 4.5 m in dry soil conditions.

If we do not consider the possibility of the ground being wet, this appears on the face of it to be adequate for the full survey. If, however, the ground is wet and the transmission velocity is, say, 0.06 m/ns rather than 0.1 m/ns, then the maximum probing depth is only 2.7 m. This is derived by applying exactly the same calculation as for dry soil and the reader is advised to check this calculation for herself/himself before continuing.

While this could potentially allow detection of part of the line of the sewer, it certainly will be incapable of detecting anything lying at a depth of 3 m or below. A transmission velocity of 0.06 m/ns is not unusual for a very wet inorganic soil. It is therefore very important to understand the implications of moisture content on the transmission speed of the electromagnetic pulses.

Had the field contained an organic soil such as peat, the problem could potentially be even worse. Peat is composed of a mixture of decomposing organic materials and water. Although conditions can vary, an organic soil has greater potential for water retention, resulting in a much lower transmission speed. In this case, a transmission speed as low as 0.035 m/ns is possible although the velocity depends directly upon the degree of waterlogging, effectively the ratio of the volume of water to that of the organic material present (Ulriksen, 1982; Utsi, 2006). If we assume the worst-case scenario of 0.035 m/ns as the transmission velocity, our calculation becomes:

$$2 \times \text{depth} = 0.035 \text{ m/ns} \times 90 \text{ ns}.$$

This gives a measure of 3.15 m for twice the depth which leaves us with a maximum probing depth of a little more than 1.5 m. The equipment we have to hand would be useless in this case. It is not, however, the GPR or GPR technology that is the problem. The operator needs to be aware of the effect of water on the speed of transmission and use a suitable radar, one with a longer wavelength and deeper probing capability. In this way, the user bridges the gap between measurement in nanoseconds time to the expected depth in meters and centimeters or feet and inches.

In practice, it is not unusual not to know the maximum probing depth capability of a radar in nanoseconds as not all manufacturers specify this on the laptop, notebook, or data logger used to control the radar. Some do, some do not. However, it is usually possible to determine this from the technical specifications of the equipment. This is particularly important when using a radar which uses the terms "shallow" and "deep" since the presence of moisture in the ground effectively changes these terms. If the readily available radar is intended for a different type of application and not suitable for moisture-rich environments, there is little point in using it for this sort of investigation. It is also not acceptable to blame water content for a lack of success in finding the targets if the radar is not suited to the site conditions.

The example given above does not take account of the potential effect of water on signal losses but this is a topic we will come back to toward the end of this chapter.

WHERE AIR MAY BE PRESENT

In a similar manner, the apparent depth may be distorted if air is present below ground. One classic example of this would be the existence of a cellar within the foundations, as yet still in place, of a building which itself has been demolished prior to redevelopment. In this case the radio waves travel very much faster through the air gap than through the adjacent soil and building materials, at 0.3 m/ns rather than 0.1 m/ns, assuming the site is dry. If the site is wet, the difference will be greater since the transmission velocity through damp ground will be less than 0.1 m/ns.

There are two implications for this. The first is that the gap where the cellar is present will look very much smaller than it is in reality because the electromagnetic pulses travel so much faster through the air gap than they do through the surrounding soil and building remains. Assume that the site has dry soil and that the transmission velocity of 0.1 m/ns is therefore a reasonable approximation of the transmission velocity through the majority of the site. Assume also that the cellar occupies the first 2 m of the subsurface and that we are probing to a depth of 2 m. Applying the same depth calculations as we did to the previous examples of wet sites gives:

$$2 \times \text{depth} = 4 \text{ m} = 0.1 \text{ m/ns} \times T \quad T = 4 \div 0.1 = 40 \text{ ns}$$

A depth of 2 m through dry soil/building materials would therefore measure 40 ns in the radargram. Repeating the identical calculation through an air gap, $T = 4 \div 0.3$ or 13.3 ns. So, within the same radargram, one area has been probed to a depth of 2 m but, in the vicinity of the air gap, the 2 m depth of the cellar is depicted at 13.3 ns. Since we are probing to a depth of 40 ns, this leaves a further 26.7 ns of depth of data generated from below the cellar floor. The 40 ns is a constant across the radargram. The physical depth is not and the depth distortion has to be taken into account by the GPR surveyor.

It is not currently possible to depict radargrams at variable transmission velocities during data collection and, although it is possible to create a velocity file which will translate the data to appropriate depths, the accuracy of this depends to some extent on how easily a representative model can be created. If the data are displayed at a translation from nanoseconds into meters of 0.1 m/ns, this will appear to show a narrow band for the cellar reaching to only 13.3 ns. At this velocity, this risks being interpreted as 66.5 cm which is an unlikely height for a cellar. The calculation of this false depth is as above:

$$2 \times \text{depth} = 0.1 \text{ m/ns (velocity)} \times 13.3 \text{ ns (probing depth in time)}$$

or 1.33 m giving a depth of 0.665 m.

Once the possibility of an air gap is recognized, then the depth can be translated at the correct transmission velocity of 0.3 m/ns and recognized as 2m. The reader is advised to carry out this calculation for himself/herself.

Now consider another example, as shown in Fig. 4.1. The feature of interest lies close to the surface between 16 and 18 m along the distance axis. There is a distinctive layer signal above and another below, marking the boundaries of a relatively homogeneous area. We cannot know what material is contained within this feature unless we carry out an intrusive test of some kind. If we were given the information that there might be a cellar directly below the floor at this point, we might reject the idea on the basis that its vertical dimension appears too shallow. Fig. 4.1 illustrates the difference in depth measurement when the transmission velocity is changed from 0.1 to 0.3 m/ns. In Fig. 4.1A, the internal height of this feature (reading from the depth axis) appears to be approximately 0.5 m at its maximum. In Fig. 4.1B, changing the transmission velocity to 0.3 m/ns gives a much larger height dimension (maximum c. 1.5 m). The height measurements are taken at the zero crossings above and below the feature. Note that neither velocity is correct for the full length of the radargram since the transmission velocity in part of the survey area is different from the remainder of the site.

In practice, as with Fig. 4.1, a void would normally have a top covering of some sort so that the real depth at any point would have to be worked out based on so many nanoseconds at one velocity (e.g., 0.1 m/ns) then so many nanoseconds for the air gap at 0.3 m/ns plus any further nanoseconds below the void at an appropriate velocity (e.g., 0.1 m/ns). A velocity file would have to be created to correct the full data set.

The second implication arises directly from this observation. This is that, in any survey line where air is present for part of the line only, the actual depth reached will not be the same throughout the survey line. The depth

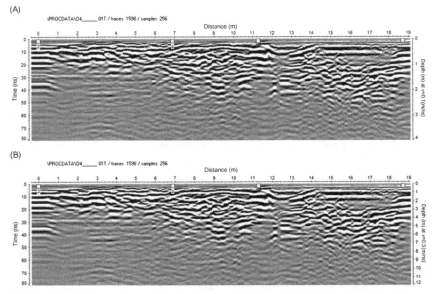

FIGURE 4.1   (A) Data displayed at a velocity of 0.1 m/ns. (B) The same data displayed at a velocity of 0.3 m/ns.

will be greater below the void than for the rest of the line. So, although the probing depth is the same time (the same number of nanoseconds), the actual physical depth will be greater where there is an air pocket.

This contrasts with the situation where water is present in one location but not throughout the survey line. The actual depth reached will be less beneath the area containing water than the probing time in nanoseconds suggests. This should not be confused with the idea that the radar is an inaccurate tool. On the contrary, the radar reads very accurately indeed in nanoseconds time. If there is an inaccuracy, it is likely to lie in the transmission velocity used to translate the time readings into depths in meters/centimeters or feet/inches.

EXAMPLES OF AIR GAPS

The above examples used a single feature, a cellar, for illustration purposes but there are many other circumstances where the same principles apply due to air gaps forming in the subsurface. Typically buried air gaps occur where there are subsurface features which had a former practical use such as mine shafts, tunnels, and cellars, or where there are defects such as delamination between layers in flexible pavement, e.g., between the wearing course and the layer beneath. Voids develop as defects in concrete structures, in historic buildings, particularly those subject to external movement such as earthquakes, build up in graves, particularly those with a permanent outer structure (e.g., stone rather than wood) and, on a larger scale, within geological contexts with the formation of sink holes or fissures in ice being obvious examples.

Depending on the dimensions of the void and the wavelength of the radar used for detection, it is not unusual to find ringing (echo effects) from a subsurface air gap which can often be a clue to the presence of the void. Fig. 4.2 shows an example of two possible voids below the floor of a historic building. The signal response close to the surface is visibly different at 1−1.8 m along the survey line and also at 3 m to nearly 4 m. The large box-like signals indicate the presence of a material, the top and bottom of which cannot be separately defined by this frequency of radar. The signal block is a combination of the signal into the material and the signal back out of the material into the environment. If it was important to know the dimensions of the feature, a higher frequency radar would be required. It is also clear that the subsurface signals are different from the rest of the data at these two positions. Below the first position there is a clear column of echo effects (ringing) which originates in the left-hand side of the near surface structure. This is a typical result of surveying across an air gap. A similar effect can be found, for instance, when surveying across a capped mine shaft. This, taken together with the strength of these signals, strongly suggests that the feature immediately below the floor is a cavity. We can be reasonably sure that

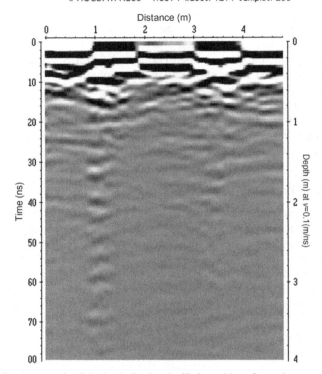

FIGURE 4.2 An example of ringing indicating the likely position of two air gaps.

the ringing is not from a metal object because radio waves cannot penetrate metal so that the echo effects would be more likely to obscure the whole of the depth below, rather than repeating at regular intervals. As metal is conductive, the echoes would also be stronger (giving a stronger contrast in the image).

On the right-hand side, there is a similar column of ringing which appears to be reflected from both sides of the subsurface structure. The strength (amplitude) of the signals is less than in the first feature. However, this also strongly suggests a second cavity below the floor of the building.

EXAMPLES OF SUBSURFACE WATER

The most common example of water affecting a survey is the presence of groundwater within the probing depth. Although most soils beneath man-made surfaces, such as asphalt or concrete, are usually dry, this is not always the case, particularly if either there is a leaking water pipe in the vicinity or rainwater is causing deterioration of either the surface or the subsurface

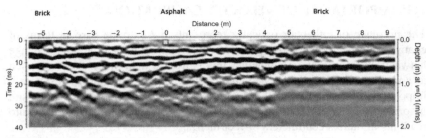

FIGURE 4.3 An example of clues to the presence of water in the subsurface.

structures. It is also not uncommon, particularly in countries with a heavy rainfall, to find that a particular area of ground has a velocity profile where the water content increases with increasing depth. This results in slower transmission velocities as the water content of the soil increases. Voids are often the result of water flow and water, rather than air, may be associated with the types of defects noted above.

It should not be assumed that the subsurface of buildings will always be dry as this depends on the soil on which they have been built, the condition of any substructures such as drains or even graves, the way in which groundwater is flowing through the ground and on the methods, and materials used in constructing the building.

It is sometimes possible to detect the presence of increasing amounts of water from the patterning in the two-dimensional GPR data because of the effect on the transmission velocity. If the volume of water present is sufficient to materially decrease the transmission velocity, the typical black and white banding of signal returns may increase in size relative to the time axis. Fig. 4.3 shows an example from a survey carried out to determine the position of a rising water main and other utilities in its vicinity. In Fig. 4.3, there are two clues to the existence of water in the subsurface. As noted on the image, the surface materials change. The central section of the radargram is covered by asphalt but both right and left sides have brick paving at the surface. On the right-hand side of the radargram, the black and white banding indicating a layer change at c. 60 cm depth appears slightly stretched in the vertical direction relative to the remaining layer signals across the data. Below this layer change, in the same area, there is also an obvious loss of signal which suggests that there is an additional attenuating material present. Water is a contributing factor to signal loss (attenuation) and it therefore appears very likely that there is moisture present in this area. This is particularly evident when we compare the right-hand side of the radargram with that on the left, also covered by brick. On the left-hand side, there is neither any sign of the signal banding lengthening (in effect, slowing down) nor is there any loss of signal in the lower subsurface. In fact, the rising main is visible as a large hyperbola on the left-hand side.

THE IMPORTANCE OF VELOCITY CALIBRATION

The presence of either air or water represents the extreme case in terms of changes in materials below ground. More common is a gradual change between different materials whether this is from a man-made structure, such as a road or bridge deck, or open countryside. This means that it is extremely important to be able to calculate the transmission velocity on any given site in order to be able to translate accurately the depth in nanoseconds into readings in meters/centimeters or feet/inches.

It is not possible to estimate transmission velocity from examination of the ground surface. Water at surface level does not necessarily indicate a fully waterlogged site. This depends on the drainage pattern, how recently the rain fell, how heavy the rainfall was, and the take-up or lack of take-up by any surface and surrounding vegetation. Equally an apparently dry surface may simply indicate a lack of recent rain rather than a dry subsurface. It depends on the drainage qualities of the subsurface materials. Measurements of relative permittivity (Er) have been used to measure transmission velocities across several similar wetland sites for which the environment and general site appearance were sufficiently alike that a reasonable guess would have estimated a similar order of magnitude for each site's transmission velocity. However, this revealed a series of very different transmission velocities even among the group of sites which proved to be fully waterlogged (see Utsi, 2007). The principal methods of velocity calibration, including the wide angle reflection and refraction (WARR) calibration used in this study, are described in the next chapter.

One major difference between the effects of air and water in the subsurface, touched on above, is the associated loss of signal into the environment. Air is not a particularly attenuating (or lossy) environment but wet soil may be, depending primarily on the free ion content but also the volume of moisture in the soil (see also Chapter 6: Attenuation or Loss). It may be necessary to use a compensation strategy which is to deploy a lower frequency antenna than would normally be required. The logic behind this is straightforward. As described in Chapter 2, Wavelengths and Why They Matter, the wavelengths emitted by the radar depend on the transmission velocity according to the formula $\lambda = v/f$. It follows therefore that if the velocity decreases, then the wavelength will shorten. Both for the purposes of reaching the full probing depth required and for identifying the survey targets, it makes sense to use a lower frequency of antenna.

Take as an example, this comparison of depth and target detection capability. In a dry soil, where the transmission velocity is 0.1 m/ns, a 400 MHz antenna will emit a range of wavelengths around [0.1 m/ns ÷ 0.4 GHz], applying the wavelength formula. This equates to a wavelength of approximately 0.25 m (250 mm). In the same environment, a 250 MHz antenna will emit a range of wavelengths around [0.1 m/ns ÷ 0.25 GHz], applying the same formula, or 0.4 m (400 mm).

If we now change the environment to one for which the transmission velocity has been calibrated to 0.065 m/ns, the wavelengths shorten to [0.065/0.4] or 162.5 mm for the 400 MHz antenna and [0.065/0.25] or 260 mm for the 250 MHz antenna. It makes sense to use the lower frequency antenna in this environment.

It should also be apparent from the foregoing that it is important to have a measure of the transmission velocity on any survey site for the depth readings to be accurate. It may be more critical in any country where sudden or prolonged rainfall is a feature of the weather pattern. The full effect of the water on wavelength will depend on how wet the site is and possibly also at what depth groundwater is present.

PRACTICE CALCULATIONS

1. What is the wavelength of a 500 MHz antenna in dry soil? And in soil for which the measured Er is 25? What is the maximum theoretical depth it can reach?
2. Is the wavelength of a 300 MHz antenna the same in a dry soil as it is in the same soil after a week's worth of rain? Assume a transmission velocity of 0.08 m/ns for the soil after the rain.
3. What is the wavelength of a 1.5 GHz antenna in dry soil? And in concrete for which the calibrated transmission velocity is 0.122 m/ns? What is the maximum theoretical depth it can reach?
4. Calculate the minimum distance between two similar pipes for both to be detectable using a 1.5 GHz antenna. The soil is damp with a relative permittivity of 16.
5. Your survey area is covered in reinforced concrete. You are told that the mesh is 200 mm by 200 mm. The relative permittivity of the concrete has been estimated at 4.5. You have three antennas: 500 MHz, 1 GHz, and 2 GHz. Which, if any, of these antennas can be used on site to detect and map subsurface targets?
6. Using a high-frequency antenna, you measure a 2 ns delamination gap in the roof of a tunnel. What is the size of this gap if (a) it is a void or (b) it contains water?

ANSWERS TO CALCULATIONS

1. Wavelength, $\lambda = v/f = 0.1/0.5 = 0.2$ m (200 mm). An Er of 25 is equivalent to a velocity of 0.3/5 or 0.06 m/ns. Wavelength = 0.06/0.5 = 0.12 (120mm). The theoretical maximum depth of 20 wavelengths would be 2.4 m.

2. Wavelength, $\lambda = v/f$. For dry soil this will be 0.1/0.3 or 0.333 m (333 mm). For the wet soil, this will be 0.08/0.3 or 0.267 m (267 mm). The wavelength will not be the same, it shortens in wet ground.

3. In dry soil we can assume a transmission velocity of 0.1 m/ns. Wavelength, $\lambda = v/f = 0.1/1.5 = 0.0667$ m (66.7 mm). In concrete, $\lambda = 0.122/1.5 = 0.081$ m (81 mm). The maximum theoretical depth is 20 wavelengths, so 1.3 m for the dry soil and 1.62 m for the concrete. The real probing depth achieved is likely to be much less than this due to signal attenuation.

4. The minimum distance between two pipes for both to be detected is 1 wavelength. First calculate the transmission velocity. $V = c \div \sqrt{Er} = 0.3/\sqrt{16} = 0.3/4 = 0.075$ m/ns. Wavelength $(\lambda) = v/f = 0.075/1.5 = 0.05$ m (50 mm).

5. Start by calculating transmission velocity. $V = c \div \sqrt{Er} = 0.3/\sqrt{4.5} = 0.141$ m/ns.

   For a 500 MHz antenna at 0.141 m/ns, $\lambda = 0.141/0.5$ or 0.282 m (282 mm). As this wavelength is greater than the mesh spacing, this antenna cannot and should not be used.

   For a 1 GHz antenna at 0.141 m/ns, $\lambda = 0.141/1$ or 0.14 m (140 mm). As the wavelength is less than the mesh spacing, this antenna can be used.

   For a 2 GHz antenna at 0.141 m/ns, $\lambda = 0.141/2$ or 0.07 m (70 mm). As the wavelength is less than the mesh spacing, this antenna can be used. The choice of antenna between the 1 and 2 will depend on the depth of penetration required.

6. If the gap is a void, the transmission velocity is 0.3 m/ns. Taking the two-way travel time into account, 2 ns measurement represents a vertical height of half of $[2 \times 0.3]$ m giving a gap of 300 mm. If the gap contains water, then the transmission velocity is 0.033 m/ns and 2 ns height becomes 0.033 m or 33 mm. Quite a difference!

REFERENCES

Ulriksen, P., 1982. Application of impulse radar to civil engineering, Doctoral Thesis. Lund University of Technology, Lund, Sweden.

Ulriksen, P., 1983. Some applications of impulse radar. Geoexploration 21, 289.

Utsi, E., 2006. Sinking into old ground: ground probing radar in the Scottish wetlands. In: Jones, R.E., Sharpe, L. (Eds.), Going Over Old Ground": Perspectives on Archaeological Geophysical and Geochemical Survey in Scotland,. University of Glasgow, Scotland, pp. 175–180. (BAR British Series 416).

Utsi, E, 2007. Wetlands viewed through the antennas of a ground penetrating radar. In: Barber, J. et al. (Eds.), Archaeology from the Wetlands: Recent Perspectives, Proceedings of the 11th WARP Conference, Society of Antiquaries, Edinburgh, pp. 213–220.

Chapter 5

Velocity Calibration

Chapter Outline

In the last chapter, we saw that, in order to make full use of the accuracy of depth readings when using a GPR, it was essential to have a measurement of the transmission velocity and, in many cases, an indication of its variability within any given site. Without this, the radar output will be accurate in terms of nanoseconds time but not necessarily so in terms of meters/centimeters or feet/inches. Since this latter information dictates what use will be made of the radar data, including the possibility of excavation in the vicinity of the survey targets, accuracy is important for the sake of safety. A relaxed attitude to an assumed transmission velocity would therefore really not be appropriate.

We also saw how a change in velocity affects the wavelength. Since the ability to detect any target is heavily dependent on using an appropriate wavelength (Chapter 2: Wavelengths and Why They Matter), it is important to have at least an indication of velocity and also whether it is variable across the site or not. In this way velocity calibration allows the GPR operator to make sure that they are using the correct GPR equipment for the survey and the site conditions.

There are at least five basic methods of calibrating transmission velocity, each of which depends, in some measure, on comparing known measurements with the GPR data. The most appropriate one for any particular survey will depend in part on the site environment and in part on the type of survey to be carried out as well as the type of equipment available to complete the survey.

CALIBRATION BY CURVE FITTING

In Chapter 1, Introduction to GPR, it was seen that the beam emitted by any GPR forms the shape of a cone. The hyperbolas seen in the GPR data are formed from the way in which that cone meets the edge of the buried targets

Ground Penetrating Radar. DOI: http://dx.doi.org/10.1016/B978-0-08-102216-0.00005-9

before passing over and beyond them. The shape of each hyperbola is formed primarily from two components, namely the radar design, specifically the shape (and angle) of the radar beam, and the electromagnetic responses of the subsurface environment, in particular, the transmission velocity of the electromagnetic pulses through the ground. The majority of the radar design element is a constant, the main exception being that, since the width of the beam increases with depth, the outline of the curve will also change with depth. The transmission velocity depends entirely on the site conditions. The shape of each individual hyperbola is therefore a function of depth and transmission velocity. This means that the curved shape of the hyperbolas in the radar data can be used to determine the transmission velocity provided that the depth is known. This is done using analytical software by fitting a digital curve to the outline of one or more hyperbolas in the data. The analytical software will automatically change the shape of the hyperbola in function of the depth.

All GPR interpretative software packages have the capability to curve fit a hyperbolic shape to the outline of a hyperbola visible in the GPR data. Fig. 5.1 shows one example of how this is done. A high-frequency (1.5 GHz) antenna was used to examine the condition of a concrete tunnel lining. Fig. 5.1 shows the same data set twice over so that the fitted curve on the data can be compared with the original data. The cursor is set to the hyperbola function. The shape of this hyperbola changes automatically with depth (following the characteristics of the cone-shaped beam which also broadens with depth). The cursor is placed over the hyperbolic signal in the data at a

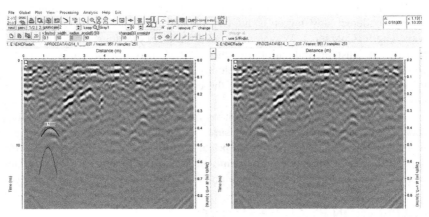

FIGURE 5.1 Curve fitting in a concrete lined tunnel. The mobile cursor (bottom left) is moved into position on the zero crossing of each hyperbola to be measured and the velocity (v) altered until a good fit is obtained.

zero crossing point. The zero crossing occurs at the interface between the positive and negative parts of the signal. This is the first interface between the black and white banding. It does not matter whether this is black to white or white to black, the curve should be fitted to the dividing line between the two. The value of transmission velocity is adjusted until the two hyperbolas follow the same outline (see Fig. 5.1).

Subsurface targets which are small relative to the wavelengths transmitted by the radar (but not less than 10% of the wavelength) will be detectable but their size is not normally measurable. These are called point targets because they act as if, from the perspective of the radar, they were a point in the subsurface, even though this may not be their physical reality. Point targets are ideal for velocity calibration because they do not reflect the size of the target, only the velocity and the shape of the radar beam.

If the target is below 10% of the wavelength, then it may not be detectable (see also Chapter 2: Wavelengths and Why They Matter). Where the target is large relative to the wavelengths transmitted by the GPR, the target may have a measurable size. For larger targets, it will be necessary to adjust the radius of the hyperbola to allow for the size of the object as well as adjusting the transmission velocity. Without the additional adjustment of the size, it would not be possible to fit the cursor to the data. In Fig. 5.1, the fitted curve shows a velocity of 0.1 m/ns which is consistent with the dry condition of the tunnel being investigated at this particular location. The size adjustment in this example is carried out by changing the radius. In Fig. 5.1 no such adjustment is necessary indicating that this is a point target (potentially a void in the concrete) but not one which is large relative to the wavelength of the radar being used. Had the target been larger, an adjustment to the radius would have been necessary in order to fit the cursor to the full shape of the hyperbola.

Although this adjustment for size sounds like a complication, in practical terms it is not. Curve fitting is a rapid process whether or not the radius or any of the other potential parameters requires adjustment. The information gained from carrying out the calibration exercise may also be a useful indicator of the proportion of targets of significant size relative to the wavelength which may, in itself, be highly relevant to the survey. Another good example of where this process gives useful information is in an archeological survey of the interior of a building where the subsurface contains large building remains such as a vault or crypt. In this instance the exercise is very useful as the dimensions of the underlying features can be estimated.

As Fig. 5.1 also indicates, there may be a need to adjust for the target lying at an angle other than 90 degrees to the path of the radar (see also Chapter 10: Three-Dimensional Data). Once the hyperbolic cursor has been activated, it is generally evident from attempting to fit the digital curve that another adjustment may be required.

Although this is a very common method of velocity calibration, it is subject to two potential flaws. The first is that the data must contain several sufficiently clear hyperbolas or it will not be possible to use this method. A lack of hyperbolas or too much clutter in the data can make it impossible to curve fit. The other potential problem is that there is a certain amount of leeway in fitting an appropriate velocity. In other words, there is usually a small range of values which will fit the same hyperbola outline. This latter problem, however, can be refined to greater accuracy by the use of a mathematical technique known as migration.

MIGRATION

Migration is a form of mathematical processing whose primary purpose is to define linear objects (e.g., pipes and services) more precisely. Correctly applied, the process collapses hyperbolas into a point within the GPR data. For linear objects, this makes their position and depth much easier to define because there is no longer a need to look at the full hyperbola (see Fig. 5.2). Although migration is a mathematical process, it is not necessary to either know or apply advanced mathematics in order to migrate data. All GPR interpretative software packages include migration as a processing step and there is a wide variety of different migration methods to choose from. The primary parameter required to collapse the hyperbola into a point is the transmission velocity. This makes the migration process one way of increasing the accuracy of curve fitting since the process only works correctly if the transmission velocity applied is accurate.

If the velocity has been accurately determined, the migrated hyperbola forms a neat point within the GPR data. Using this point rather than the full hyperbola means that the position of the subsurface target can be more easily and more accurately determined. However, if the transmission velocity applied in the migration process is too fast, then the collapsed hyperbola will appear as a slight upward curve rather than as a central point. If this happens,

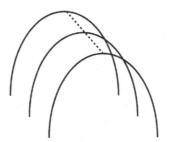

FIGURE 5.2   Migration collapses hyperbolas to a point. The image shows hyperbolas in three successive survey lines.

reduce the transmission velocity and reapply migration until either a single point appears or a downward curve appears. If no curve is present, then the velocity used to migrate the data is the accurately calibrated velocity. If a downward curve appears, then the velocity used is too slow and the migration process should be repeated at a slightly increased velocity.

Similarly, if the initial velocity applied were too slow, the collapsed hyperbola will appear as to have a slight downward curve rather than being a point. Increasing the velocity used to migrate the data will remove this curve. Once upward or downward curves no longer appear in the migrated data, the velocity used for migration is a more accurate calibration.

Fig. 5.3 illustrates the effect of applying migration to two hyperbolas. Interestingly there appears to be a difference in transmission velocity at the two positions since the lower hyperbola forms a neat dot indicating that the correct transmission velocity has been used whereas the upper hyperbola has a slight upward curve, indicating that the transmission velocity used was too high. If the velocity is slightly less at this point, the most likely cause of is a higher moisture content in the immediate vicinity.

In Fig. 5.4a lower velocity has been applied to the same data. The upper curve now appears as a discrete dot indicating confirmation for this velocity at this position. The lower curve now slopes downward, indicating that the velocity at this position is higher than that used for the migration. In the right-hand trace, a faster velocity has been applied and the signals for both hyperbolas have a slight upward curve, indicating that this velocity is too high.

Both curve fitting and migration are steps taken after the data have been recorded and these methods of velocity calibration are therefore usually applied after the survey (see also Chapter 8: Data Processing).

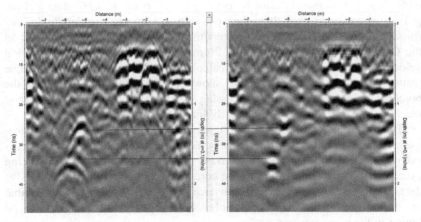

FIGURE 5.3 The effect of applying migration—unmigrated data to the left, migrated data to the right.

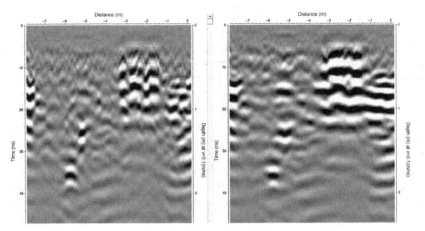

FIGURE 5.4 Migration at different velocities—slower (lhs) and faster (rhs).

KNOWN LAYER OR LAYERS

For this method of calibration, it is necessary to know the depth of at least one subsurface layer accurately, in one or more locations. It is commonly used on man-made layered materials, e.g. in road, bridge deck, and airport investigations. It is also a common method of calibrating velocity in environmental investigations. For multiple layers, each layer reading can be used to calibrate the transmission velocity within that particular material. The depth readings in each case are taken by extracting cores from the survey site at predefined positions.

For this calibration to work effectively, the position of the boreholes from which cores have been extracted must be accurately identifiable on the GPR data. This is usually done by placing a marker on the GPR data and the ground at the same time. It is then possible to compare the depths derived from the extracted cores in centimeters/meters (or feet/inches) with those in the GPR data in nanoseconds time. From this comparison, the transmission velocity of the radio waves can be calculated. It is then a relatively simple matter to adjust the physical depth shown on the GPR data to that of the borehole data, using the postprocessing analytical software. Alternatively, the velocity reading can be fed directly into this software package.

The most common problem with this method lies in the timing and placement of the boreholes and the markers. For this method to be accurate, it is essential that the GPR survey is carried out in advance of the cores being extracted and that the location of the cores is accurately marked on the GPR data. If the cores are extracted in advance of the GPR work, it is impossible to survey the location of the boreholes since the relevant materials are no longer in the same position.

Where cores have been extracted in advance of the GPR survey, it is common practice to compare the contents of the borehole with the GPR data collected from a nearby location. Unfortunately this is only accurate if the subsurface layers lie at a constant depth across the survey area. Even in built environments, this is far less likely than is sometimes assumed. Layers of asphalt and concrete frequently vary over relatively short distances, making the comparison inaccurate and the calibration invalid. In open countryside, where the stratigraphy has been laid down by natural forces and there is no requirement for leveling, it is frequently highly inaccurate. An inaccurate depth reading resulting in an inaccurately calculated transmission velocity will result in depth errors for the whole survey. This is not inaccuracy on the part of the radar.

Where coring is carried out at the same time as the survey, as is usual practice on environmental surveys, it is generally possible to coordinate the positions on the ground and on the GPR data. In the built environment this is less likely to be practical. It is not unusual for two independent companies to be carrying out the GPR survey and the borehole extraction and they may not be on site at the same time. Assuming that the survey is carried out in advance of the cores being extracted, the GPR survey team should mark the positions on the ground where they require cores to be taken as well as placing markers on the GPR data in the corresponding positions. The coring team should be briefed to extract cores at these locations so that a valid comparison can be made and an accurate transmission velocity be applied to the full GPR data set. If this does not happen, there is a risk that the calibration will not be as accurate as it could or should be. Siting boreholes and the GPR calibration points in different locations has the potential to introduce depth inaccuracies into the GPR results.

There is another potential problem which can occur in open ground. Unlike layers of construction materials, buried soils do not always correspond visually to their electromagnetic composition. A person extracting a core will categories layer change on the basis of visual information which often includes a measure of particle size, color, and similar parameters. The radar can only detect changes in electromagnetic response. It is not unusual, for example, in the United Kingdom, to have a borehole log which reads "sandy clay, clayey sand" with color descriptions attached. Unless these descriptions correspond to measureable changes in electromagnetic response, it may be difficult to reconcile the borehole log with the radar data. There is a risk that the major change in electromagnetic properties does not coincide with the perceived layer change. The comparison process is obviously very much easier where the stratigraphic changes relate to very different materials such as peat to granite or a sandy soil to waterlogged clay, for example.

OBJECT AT KNOWN DEPTH

This method relies on a specific target being already in place on the site to be investigated or on being able to place such a target accurately within the survey site environment. As this is a relatively unusual circumstance, the method is most commonly used on test sites where a known target has been built in as a calibration feature and for which an accurate depth has been recorded.

As with comparison with a layer at a known depth, it is necessary to be able to accurately locate the point at which measurement is taken over the object both physically on the ground and digitally on the radar data. Comparison of the depth reading in nanoseconds can then be made with the depth reading in meters/centimeters or feet/inches, deriving the transmission velocity as a result. This can be done either immediately in advance of survey, during the survey, or immediately afterward. It should not be done on a day other than the survey day since transmission velocity will vary with the weather on outdoor sites and may vary if there is any variation on moisture content or the level of groundwater on indoor ones. Some GPR equipment allows the operator to make this calibration adjustment on-screen at the time of the survey rather than having to wait until the postprocessing stage.

The main problem with this method of calibration is similar to those from using a stratigraphic layer at known depth. If the data are extracted physically and digitally from different locations, there is a risk of an inaccurate calibration. In addition, there is a risk, even with test sites, of movement caused by drying out or settlement from repeated use and on external sites, animal activity, e.g., rabbits, can disturb the test object from its accurately calculated position.

WIDE ANGLE REFLECTION AND REFRACTION

Wide angle reflection and refraction (WARR) is also sometimes called common midpoint or CMP for short. This method of velocity calibration relies on having separable antennas. It is therefore most frequently used in environmental investigations and often with lower frequency antennas. The object of the exercise is to compare the direct signal through the air from the transmitter to the receiver with that of a simultaneous signal through the ground. Since the velocity of the signal in air is known (the speed of light, or approximately 0.3 m/ns), it is possible to use a simple computer program to compare the two signals and, from the comparison, work out the velocity of the signal through the ground.

The calibration is carried out by placing a receiver antenna immediately adjacent to a separately boxed transmitter antenna on the ground. One antenna is then drawn away from the other while the radar continues to transmit. It does not matter which antenna remains stationary and which is drawn away. The sampling interval (or time between readings) should be set to a minimum in order to get a reasonable number of readings defining the curved shape of the signal. Transmission in continuous mode is, however, not advisable. As the antennas are drawn apart, a minimum of two layer-type traces should appear on screen, that of the air signal and at least one through the ground. The precise number will depend on the number of underlying stratigraphic layers. The uppermost signal will be the direct signal from transmitter to receiver. Curve fitting is used to match this signal to the known transmission velocity of radio waves through air and to obtain a reading for Tzero, the time for the signal to travel down the cables before transmission into the ground (see also Chapter 3: More Fundamentals of GPR Operation). A process of altering the two unknowns, depth and velocity (or Er, the relative permittivity), is then used to match the line of the next curve. This either gives a direct reading of the transmission velocity or of the relative permittivity, depending on the type of processing software used.

If the reading is of direct permittivity, the velocity can be calculated from the formula:

$$v = c \div \sqrt{Er}$$

where v is the transmission velocity, c is the speed of light (approximately 0.3 m/ns), and Er is the relative permittivity. This process can be repeated for lower layers, if present. Some systems have dedicated in-house packages for determining the transmission velocity based on the frequency of antenna used and the sampling parameters used. Fig. 5.5 shows an example of the curve fitting process for WARR. The top signal is the direct

FIGURE 5.5   An example of WARR calibration.

signal from transmitter to receiver and the lower signal the return from a subsurface layer.

One of the basic assumptions of this method is that the subsurface layer being surveyed should be flat or reasonably flat which makes this method ideal for the survey of the middle section of peat basins, for example.

TRANSMITTER WITH MULTIPLE RECEIVERS

This is an extremely efficient method of calibrating velocity but it is also one of the least common. It requires the use of a single transmitter paired with a minimum of two receivers, preferably three or four. It also requires that all of the antennas are simultaneously triggered, i.e., there is no time delay built into the system and the transmitter and the receivers are all triggered at the same time. A very small time delay is built into most (but not all) GPR systems in order to avoid cross channel interference. Before the method is applied, it is important to measure the Tzero accurately on each receiver so that the cable travel time can be eliminated from the calculation. It is also important that there is no cross channel interference as this would obscure the layer data used to compare the individual channels. An individual channel in this case is the transmitter plus any one of the receivers. For example, if three receivers are used, there will be a total of three channels in operation, being Tx to Rx1, Tx to Rx2, and Tx to Rx3. In a similar manner, should four receivers be used, then there will be four channels or for only two receivers, two channels.

The most common use of this calibration method is in road condition surveys where there are easily identified layers which can be compared between the individual channels. The great advantage of the method is the continuity of the calculation over the length of the survey and the speed with which it can be done. Depending on the antenna requirements of the survey, it may be possible to carry out calibration simultaneously with the survey. Another advantage is an increase in the level of information. Instead of having a calibrated velocity from a few specific places, this method produces data which applies to the whole of the survey coverage. This means that it can also be used to highlight areas where the velocity varies from the expected or average value. The variation in velocity potentially indicates the presence of different materials which is useful information in itself as, if it is not the result of differences in construction, then it highlights all the anomalous areas. These are likely to be either where repairs have already been completed or where repairs are required.

The antennas are placed in a line, immediately adjacent to one another, beginning with the transmitter and then each of the receivers in turn. The method relies on knowing the distances between the transmitter and each receiver. Since the source of the transmitted pulses is the same for each receiver, and each of the antennas is triggered at the same time (so that there is no delay between any of the channels), it is possible to infer the transmission velocity from comparison of the data returned to each of the receiver antennas.

Once the data have been acquired, the first channel is analyzed by selecting ("picking") the layer digitally in the interpretation software. The software package will then compare this data with that generated in the other channels. This produces a data stream of calibrated velocities for the full distance of the survey which can be viewed either graphically or as tabulated output. The corrected velocities can also be applied to the survey data so that the depths shown are automatically corrected for transmission velocity.

OTHER CALIBRATION RULES

Regardless of which method of velocity calibration is used on a survey, there should be a minimum of one calibration exercise on each site on each day. This is because the velocity may change, particularly if there is a significant alteration in the weather, e.g., heavy rainfall. If weather conditions alter during the day, it is sensible to carry out another calibration exercise to ensure that changes in transmission velocity are picked up. It is also good practice to carry out more than one calibration exercise especially where there is any indication that one area may contain more moisture than another, for example.

TABLE 5.1 Approximate Transmission Velocities and Their Equivalent Relative Permittivities

Material	Typical Transmission Velocity	Equivalent Er (Relative Permittivity)
Ice	0.15 m/ns	4
Dry sand	0.2–0.12 m/ns	2–6
Wet sand	0.095–0.055 m/ns	10–30
Concrete	0.1 m/ns	9
Asphalt	0.122 m/ns	6
Dry soil	0.1 m/ns	9
Wet inorganic soil	0.077–0.055 m/ns	15–30
Wet organic soil (e.g., peat)	0.05–0.035 m/ns	40–75
Freshwater	0.033 m/ns	80–81

CONSISTENCY WITH SITE CONDITIONS

Whichever method of velocity calibration is applied to the data, it is sensible to check that this is consistent with the site conditions. For example, it is not reasonable to use a velocity of 0.035 m/ns for the ground conditions beneath a building. This velocity is in line with reported velocities for fully water-logged peat mosses which would be difficult to walk across, let alone build on. It may, however, be a reasonable velocity for the transmission through an underground railway tunnel where groundwater has penetrated both the soil above and also the buried structure.

Table 5.1 gives some common ranges of transmission velocity. This table can be used for comparative purposes but not as a method of calibration.

Chapter 6

Attenuation or Loss

Chapter Outline

There are a number of reasons why GPR might not work or might appear not to work. It is usually possible to identify the reasons. The most common situation where GPR will not work well is in an environment containing a high level of free ions. Free in this case means with freedom to move. Typically, this could be saltwater or some (but not all) wet clays although, in practice, there are a range of other soils equally unsuited. In this type of environment, the transmitted signal passes into the ground as a weak electrical current and then dissipates. As little or, in the extreme case, none of the signal is returned to the GPR receiver antenna, there is little or no data for the GPR operator to analyze. Most, if not all, of the subsurface information is lost.

ATTENUATION

The key elements are an ionized soil (or equivalent material) and a means of mobility. If one of these factors can be removed, then using GPR becomes a possibility. Take the example of a wet clay which is not suitable for GPR survey. This can often be surveyed once the ground has dried out. Obviously the chemical composition of the soil is unchanged but removal of the water, the means of mobility, results in less lost signal and may therefore give satisfactory GPR survey results. Fig. 6.1 shows a typical example of the detection of a culvert in clay in the summer months, in this case using a 400 MHz antenna. In the winter, this target was effectively invisible.

One clue to signals being lost (or attenuated) lies in the faint signals lying below the real black and white data marking the layer change associated with the culvert (the top of the hyperbola lies immediately above). These

Ground Penetrating Radar. DOI: http://dx.doi.org/10.1016/B978-0-08-102216-0.00006-0

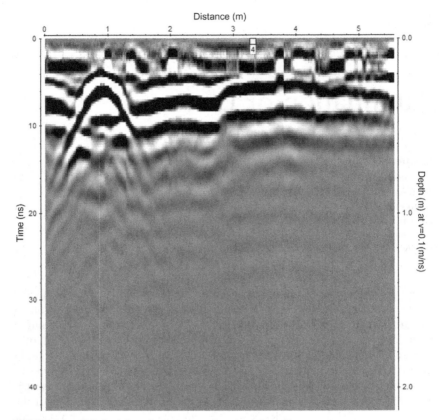

FIGURE 6.1 GPR detection of a drainage culvert in a clay soil.

follow the exact same pathway, a little lower down in the radargram, as that of the strong layer above. This shape mimicry indicates that this is an echo effect generated from the layer of the ditch above. It is typical of wet clay that little or no real signals are visible but that echo effects from shallower depths are. Even in the summer months, the layers below the ditch in which the culvert lies are not suited to GPR survey. This is likely to be because there is still moisture in the soil at these lower depths. If the subsoil were to dry out, it is possible that more real signal returns would be visible.

TOTAL ATTENUATION

It is useful to know what total attenuation (the technical term for completely lost signal) looks like. Fig. 6.2 shows an example of data collected on a waterlogged clay site in SW England using a 400 MHz antenna. The site was of archeological interest (a historic battlefield) and archaeologists were

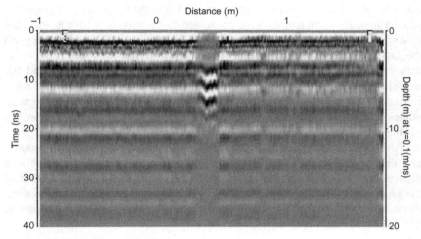

FIGURE 6.2   Total signal attenuation.

keen to survey the site, looking for any remains which could be associated with the 15th century battle. The data apparently show one large target in the middle of an otherwise homogeneous soil with suspiciously regular horizontal banding. This would be an inaccurate reading of the data.

The evenly spaced horizontal lines before and after the "target" are background echo effects. The regularity and repetition of the lines are the clues to what is happening. Lines in radar data which either mimic those directly above in shape or recur at regular intervals directly below are echo effects, not original data from subsurface targets. The constant spacing between the lines is a clue that this is not any sort of reflection of the material contained in the subsurface. What is happening in this case is that the electromagnetic pulses are passing into the wet clay of the ground, setting up a weak electric current and dissipating away from the radar. Since no signals are being returned to the receiver antenna, the full energy is being lost into the wet clay soil. The echo effects, known as ringing, are not in this case repeat reflections from the surface. The regularity of their appearance indicates that they are a function of the system design. This is, in fact, the reflection of the inherent background ringing of the GPR system. All radar systems, regardless of manufacturer, have a degree of built in echo effects. These are very weak and therefore not normally visible to the person using the radar because they are usually masked by the much stronger reflected real signals being returned to the receiver. This particular type of ringing is only visible because no such signals are being returned to the receive antenna.

What, then of the "target" in the center of the radar plot? Is it real? It looks very much like a layer signal which is exactly what it is. The ground is extremely wet and the returned signals at this point mark the position of a puddle on the surface. Since the puddle has been created by rainfall, it consists of relatively clean freshwater. The saturation of the surrounding ground appears sufficient to have prevented much mixing with the wet clay below. As there is no intrinsic problem in surveying in freshwater, beyond the fact that the radio waves will travel more slowly through water, the radar data are illustrating the interface between the water of the puddle and the wet clay underneath. Below the interface, there are no signals visible and it can safely be assumed that there is total attenuation of the electromagnetic pulses. Note that the intrinsic background ringing of the system is also not visible since the one piece of real data has effectively obscured this.

Although, at first glance, the signal from the interface between the puddle and the ground appears to come from below the surface, this is actually a depiction of the depth of the puddle combined with the slower transition of the electromagnetic pulses through the water. As can be seen from the data, there is a significant difference in transmission velocity (see also Chapter 4: The effect of Water and Air).

DEGREES OF ATTENUATION

In practice, all soils contain some elements which will attenuate the energy of the radio waves as they are transmitted through the survey environment. As discussed in Chapter 2, Wavelengths and Why They Matter, the maximum amount of energy emitted by any radar is, for most survey environments, limited to 20 wavelengths or less. In practice, it is very unlikely that a full depth of 20 wavelengths will be achieved, barring exceptional circumstances. All soils and building materials have an in-built capacity to absorb some of the signal. This is a characteristic trait of their electromagnetic properties. A soil which absorbs a high proportion of the electromagnetic pulses is said to be lossy and is less suited to GPR operation than one which propagates a higher proportion of the radar signals.

There are therefore two practical lessons to be applied from this. Firstly, in selecting a suitable antenna for any survey, it is essential to allow for a much lesser depth range than the maximum possible 20 wavelengths. Secondly, the presence or absence of groundwater at the time of the survey may be critical to the success of the survey. If the soil composition is not heavily ionized, then allowance will have to be made for the slower transmission velocity of the electromagnetic pulses (cf Chapter 4: The effect of Water and Air) but it should be possible to survey regardless of season. If the ground is either not suited to GPR survey or only partially suited. It may also be necessary to plan that the survey takes place at a time of year when the soil has dried out. This is something which can be tested in advance of survey.

It is important to take care that survey planning takes into account both the potential effects of the soil and any water content. Should a survey be unsuccessful, the correct reason for the failure needs to be clearly identified. It has been known for a GPR survey to be unsuccessful because the operator forgot to take into account the slower speed of transmission due to the water content and therefore used either the wrong antenna or too shallow a depth setting, only for the soil to be blamed as "unsuitable." Whether the problem lay with the choice of antenna or with the depth setting of the radar, there is a potential solution which was not considered and should have been. It is possible that a lower frequency antenna and/or a longer time range should have been used and could have given a positive result. Alternatively, a small sample of the soil could have been tested to prove its suitability or unsuitability. The process of analyzing the soil's electromagnetic response might indicate, for example, that the survey should have been postponed to a time when the ground was drier.

ELECTRICAL AND MAGNETIC LOSSES

One complication which is worth bearing in mind is that the electromagnetic response of any soil has two components, the electrical and the magnetic. It used to be thought that electromagnetic losses increased with frequency so that the losses from a 2.6 GHz antenna ought to be greater than those of a 1 GHz antenna. In fact, this is not the case. Recent soil research confirms that electrical losses do vary linearly with frequency but magnetic losses peak at frequencies of the order of 1 GHz and then decrease (Serbin et al., 2001; Thomas et al., 2006). The higher the antenna frequency, the greater the electrical loss component, but the magnetic losses of antennas above 1 GHz in frequency are less than those of the 1 GHz antenna. This is worth remembering because, if the results of surveying with a 900 MHz/1 GHz antenna are not as good as expected, a better result might be obtained using an antenna of 1.5 GHz or above. Obviously the required probing depth will also have a bearing on the choice of antenna frequency.

NEAR SURFACE REFLECTORS

Where there are strong reflectors close to the surface, e.g., reinforcement bars or mesh in concrete, a high proportion of the signal transmitted into the ground will be reflected directly from these features. Although this is not strictly speaking attenuation, it is signal which is lost to the lower subsurface.

A good example of this is shown in Fig. 12.1, Chapter 12, Examples of Practical Problems. The survey to locate reinforcement bars was carried out through the underside of the concrete slabs (i.e., the ceiling) because there was reinforcement mesh above. This radargram shows good signal strength for the rebars and lesser signal amplitude in the returns from the mesh. Had the survey been carried out from above rather than below, there was a risk that the reinforcement bars would not have been as clearly visible. In the case of a target which is inherently less easy to detect, such as a plastic pipe beneath reinforcement bars, the loss of signal could be more critical.

FRESHWATER

Another common misconception in the use of GPR arises from the role of water in providing the means of mobility. This is sometimes equated to "GPR could not be used in water" which is inaccurate. Provided that the water is reasonably free from ionized material, there is no barrier to using a GPR except that the slower transmission velocity must be taken into account. A depth setting suited for dry soil will need to be increased, potentially to a maximum of three times the original value (cf Chapter 4: The effect of Water and Air).

Freshwater is inherently suited to GPR survey. As an example, Fig. 6.3 shows a radar profile taken across a fast-flowing river in front of a bridge. The purpose of the survey was to look for bridge scour, i.e., the removal of the sediments around the bridge piers, a particular problem with fast-flowing rivers. Not only does the GPR penetrate the freshwater in the river, showing the sediment beneath but it has also detected the evidence of suspended sediments within the water and the signals have also penetrated the sediment bed beneath the river.

The rules for water being suitable for investigation using a GPR are the same as those for soil and building materials although the inclusion of any ionized material is clearly more critical since the mobility element is already present. Material to be wary of includes some clays, additives such as fertilizers, heavy concentrations of urea, and chemical deposits. Salt is probably the most problematic material in both wet and dry conditions since, even where soils are dry, it effectively draws in moisture. There are large areas in the hot countries of the Middle East, e.g., where GPR is difficult if not impossible to apply as a survey method due to the high concentration of salts contained within the sand dunes combined with the limited amount of moisture which holds the center of the dunes in place.

Working on coastal sites can be problematic since both wind action and periodic flooding can result in the deposition of salt within soils which would otherwise be well suited to investigation by GPR. This does not necessarily mean that it will not be possible to use GPR in these areas. It may

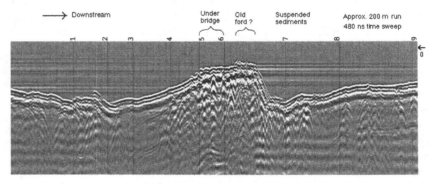

FIGURE 6.3 Radar data from a fast-flowing river showing the variable outline of the riverbed, suspended sediments, and some of the material below the riverbed.

simply mean that the depth that the radar can penetrate is restricted or that the survey would be better done during or after a prolonged dry spell. Since attenuation decreases with frequency, it may also be an advantage to use a lower frequency of antenna than one would normally in order to compensate for the signal losses as well as improving the potential depth penetration. How much this improves matters is probably best examined by a short trial in advance of the survey.

SOIL ANALYSIS

It is possible to analyze soil or water in advance of survey to determine whether or not it is suitable for investigation using a GPR. A very small sample is all that is required, say 50 cm^3 per test but this may also depend on the testing service so should be checked before submitting a sample. The soil or water sample is tested using a network analyzer to measure real and imaginary electrical impedance at a range of frequencies. The precise range of frequencies chosen will depend on the type of GPR survey which is being planned. From this data, the ratio of imaginary to real impedance can be calculated. This gives a guide to the suitability (or otherwise) of the sample for GPR survey. It also gives an indication of the most suitable antenna frequency. An example of the results of soil testing is shown in Fig. 6.4.

In this example, five soil samples have been taken at half meter intervals from the surface down to 2 m. From practical experience, the ratio Q needs to be 4 or above for GPR survey to be viable. Fig. 6.4 shows that although the surface is suited to GPR survey, increasing depth from 0.5 m downwards

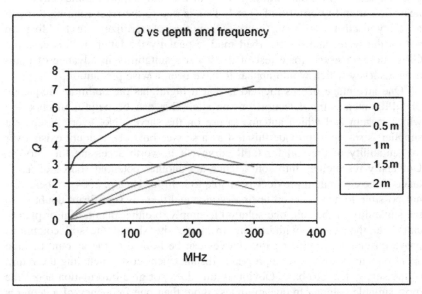

FIGURE 6.4 Results of soil testing showing decreasing suitability for GPR survey ($Q > 4$) with increasing depth.

renders each successive layer more unsuitable than the previous one above. The example in this case was a large field in an area of peatland (organic soil). As the field had previously been used for growing vegetables over many years, there has been a build-up of fertilizer in the lower reaches of the soil, rendering the field unsuited to deep survey using a GPR. In contrast, the surrounding land had been used to harvest peat for garden use and was entirely suited to peat depth assessment by GPR, even though this area also contained significant moisture (groundwater). It was the combination of fertilizer and moisture that made this field unsuitable even though, compared to the surrounding area, it was relatively dry. Where soil is being tested it is usual to test both a dry and at least one wet sample which will then give comparative results. A decision can then be made based on these results as to the timing of the survey as well as the best frequency of antenna to use.

The test procedure for water is essentially the same as that for soil or rock. Although in principle freshwater should not present a problem for GPR it is worth noting that freshwater within agricultural areas can sometimes suffer from an accumulation of fertilizer, e.g., where the surrounding agricultural land drains into it and heavy applications of fertilizer are made to the crops around it.

Fig. 6.5 demonstrates this problem. Three samples of water and one of riverbed mud from the same river, in principle therefore freshwater, have been analyzed and compared with distilled water. The results from the test site are so far removed from the entirely suitable response of the distilled water that it is almost impossible to see them on the same graph (Fig. 6.5 top). The lower part of the graph containing the test results has therefore been expanded (Fig. 6.5 bottom). None of the four samples comes anywhere near the suitability criteria of $Q = 4$. This is due to the river running through a valley used for growing crops. Quantities of fertilizer are building up within the water and the riverbed mud, especially the latter. In this case the GPR surveyor asked for a test of the water's suitability in advance of planning a survey which would appear to have been a wise precaution.

One advantage of this method is that it highlights the optimum frequency of GPR antenna for detection on the site. This can be helpful in choosing which system and which antenna to use on the survey. Not every equipment supplier is set up to provide this test as a service but, where doubt exists over the suitability of the soil for GPR survey, it is worth accessing this service. University researchers into soil science are another potential source of assistance if the equipment provider is unable to help (cf Huang, 2004). Where it is not possible to access a soil testing service and there is significant doubt as to the suitability of the soil, one solution is simply to either plan a trial or plan to carry out the survey; withdrawing from the site should the soil conditions prove unsuitable. In either case, charges can be limited to the amount of time spent on site, with, where appropriate, displacement costs, including the return of the survey team to base. Obviously this does not give information about the most suitable antenna frequency unless more than one frequency of antenna is brought to site and tested. For practical reasons, this is not always possible.

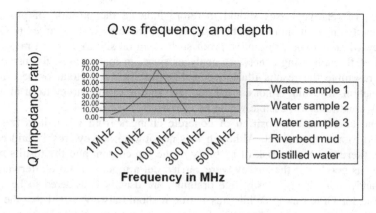

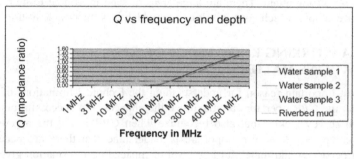

FIGURE 6.5 Water testing results for a river within agricultural land showing unsuitability for GPR investigation by comparison with a distilled water example.

It is also important that the implication of moisture and/or groundwater are taken into account, as noted above, before declaring the soil unsuitable to be sure that it is indeed the soil that is unsuitable and not either the equipment brought to site or the settings used in the survey.

FOOTNOTE ABOUT CLAY, CONCRETE, AND SAND

Clay is not one uniform substance: there are a number of different types of clay with different chemical compositions. Although all of them suffer attenuation when wet, not all of them attenuate to the same degree. There is a small body of experimental work which indicates that certain clays, when wet, can demonstrate differential degrees of signal loss. Effectively the patterning produced in the GPR data will indicate which areas are wetter and therefore lossier (cf Conyers, 2004). Depending on what the investigation targets are, this may or may not be useful. One obvious example where it could be useful is an archeological site composed of pits and ditches. These features typically retain water and will therefore be wetter than their surrounding environment. Being wetter, a greater proportion of the signal will be lost in the target areas. The pattern of relative losses should therefore indicate the position of pits and ditches. Even so, it should not be assumed that wet clay will be suited to GPR survey without some substantive proof such as an on-site trial or test measurement of the soil using a network analyzer since, in many cases, the response may resemble the results illustrated in Fig. 6.2, i.e., with total or near total attenuation. If testing is not carried out, then timing of the survey to fit in with as dry conditions as possible would be advisable.

The chemical composition of concrete changes from when it is freshly poured to after is has set. Although it is possible to survey freshly laid concrete, there is a higher risk of mobility in the ion content and the results may not be as good as if the survey were carried out a month or more afterward.

Sandy soils, being mostly free draining, are usually considered to be very suitable for GPR survey. While this is true it should also be noted that the ion content is still important. There are large areas of sand in Middle Eastern countries, for example, which contain salt and on which it is impossible to use GPR.

WHY A WORKING KNOWLEDGE OF GPR LIMITATIONS MATTERS

The most important reason for understanding the technical limitations of GPR is so that the GPR operator can maximize the chances of a successful outcome for each survey tackled but also recognize when GPR is not the appropriate investigative tool. It is also important to understand that there are occasions when a persistent and unhelpful low level of misleading information given out by unproven technology, typically claiming to overcome all possible GPR limitations, may be offered. Examples of this include no limitation on probing depth and no degree of attenuation in saltwater. Such claims should be treated with great caution. If the technology works, it can be verified in advance by a blind test in order to avoid the potential of an expensive mistake.

REFERENCES

Huang, Y., 2004. Laboratory measurements of dielectric materials. In: Daniels, D.J. (Ed.), Ground Penetrating Radar, pp. 110–129.

Serbin, O., Or, D., Blumberg, D.G., 2001. Thermodielectric effects on radar backscattering from wet soils. IEEE Transact. Geosci. Remote Sens. 39, 897–901.

Thomas, A.M., Metje, N., Rogers, C.D.F. and Chapman, D.N., 2006. Ground penetrating radar interpretation as a function of soil response complexity in utility mapping. In: The Proceedings of the 11th International Conference on Ground Penetrating Radar. Columbus, OH.

Utsi, E., 2007. Wetlands viewed through the antennas of a ground penetrating radar, In: Barber, J. et al. (Eds.), Scottish Wetland Archaeology Project. Archaeology From the Wetlands: Recent Perspectives. Proceedings of the 11th WARP Conference. Society of Antiquaries, Edinburgh. 2005, pp. 213–220.

Conyers, L.B., 2004. Moisture and soil differences as related to the spatial accuracy of gpr amplitude maps at two archeological test sites. In: Slob, E., Yarovoy, A., Rhebergen, J. (Eds.), Proceedings of the Tenth International Conference on Ground Penetrating Radar, vol. 2, 435–438.

Chapter 7

Survey Strategies

Chapter Outline

The quality of the results of a GPR survey is only as good as the survey strategy that was applied to obtaining the data. Although different applications of the technique may require different strategies, there are a number of issues which will need to be decided in advance. There are two aspects to this, the first being the technical decisions taken by the GPR surveyor in order to find the survey targets and the other being the vital interaction between the GPR surveyor and their client. In this chapter, we will deal first with the importance of survey parameters such as antenna frequency, probing depth, and sampling on the potential outcome of the survey before considering the information which anyone commissioning a GPR survey ought to be prepared to discuss.

Although it is true that a single survey line provides a vertical view along the path traced by the radar, this is far from being a full picture of the composition of the subsurface and well below the capabilities of what the radar can provide. It is important to be able to choose an appropriate level of sampling for the type of survey and also the degree of detailed information that is required from the GPR.

ANTENNA FREQUENCY

The first and most important decision in any survey is what frequency or frequencies of antenna are required. It should never be assumed that the radar which is the easiest to access is the appropriate one for the task. Nor is it necessarily safe to use a particular frequency because this is what someone

Ground Penetrating Radar. DOI: http://dx.doi.org/10.1016/B978-0-08-102216-0.00007-2

else did on a similar project. There is a risk of failure if the nature of the site differs between the previous project and the current one.

There is no substitute for obtaining as full site information as possible in advance of survey. Based on this it should be possible to work out which frequency of antenna is required. It is worth bearing in mind that more than one frequency may be required, either for different tasks or because of some uncertainty in the site characteristics, e.g., the lossiness of the soil.

As a bare minimum, the following questions should be asked of whichever radar it is proposed to use.

- Is the wavelength long enough to reach the target(s)?
- Is the wavelength short enough to be able to detect the target(s)?
- Where is the groundwater relative to the target(s)?
- What materials lie above the target(s)?
- What is the maximum probing depth required?
- What kind of soil is present and how will this affect the transmission?

For any survey where there is either a significant lack of information or a significant level of uncertainty about the information that is available, assuming that it is possible to do so, it is sensible to bring more than one frequency of antenna, if only as a backup, in case the planning assumptions prove to be incorrect.

PROBING DEPTHS

It may seem perverse to require an estimate of target depth when one of the purposes of using a radar is to determine the depth of targets. However, this is essential so that the surveyor can make sure that the GPR depth settings are adequate including both allowing a margin of error on the estimate and taking the level of groundwater or moisture content (if applicable) into account. If the depth setting is insufficient, this cannot be altered by postprocessing and there is a risk that the survey will have to be repeated.

As discussed in Chapter 4, The Effect of Water and Air, GPRs measure in nanoseconds time but this can be related to depth settings in meters/centimeters or feet/inches by translating using an estimated or actual calibrated transmission velocity. For the measurement of target depths, an estimation is not sufficient and, as discussed in Chapter 5, Velocity Calibration, a calibrated velocity must be used. For the purposes of survey planning, however, estimations are sufficient. Where the soils (or equivalent materials) are likely to be free draining and dry, the rule of thumb based on a transmission velocity of 0.1 m/ns is that 20 ns is equivalent to 1 m of depth. Following the guidance given in Chapter 4, The Effect of Water and Air, depth settings in nanoseconds can easily be calculated for different transmission speeds, allowing for possible water content in the soil. This process is neither complicated nor time consuming and it often makes the critical difference between success and failure in a GPR survey. As was illustrated in Chapter 4, The Effect of Water and Air, if the probing depth is too shallow for the transmission speed, the target(s) will simply not be detected. Failure to make this sort of calculation is at the heart of many missed targets, particularly in the field of utility detection. Where this happens, it is not a limitation of the GPR performance so much as of the survey planning process.

PRESET PROBING DEPTHS

Some systems have preset probing depths, typically those systems designed specifically for utility detection and also some of the very low-frequency radars. This feature is designed to make survey planning easier by removing the necessity for calculating probing depth. However, depending on the survey environment, it can result in either too large or too small a probing depth. In the former case, the targets will be detected but may not be as visible on-screen as one would wish. This can easily be corrected provided that the data are processed after collection since all processing packages have some form of zoom function. In the latter case, the system will be unable to detect targets on sites with high water concentrations, for example. The solution is simple. Check the technical specifications of the radar and carry out the same calculations as outlined above. This allows the operator to work out whether this radar is suitable for the task in hand or not.

EXERCISE

The answers to the following calculations are given at the end of the chapter. The reader is recommended to work out the answers for himself/herself before turning to the answers.

1. What is the calculation that links the depths of 20 ns, 1 m, and 0.1 m/ns, as per above?
2. What should the depth setting be, in nanoseconds time, for a probing depth of 2 m where the transmission velocity is 0.08 m/ns?
3. What should the depth setting be, in nanoseconds time, for a probing depth of 2 m through freshwater ice where the transmission velocity is 0.15 m/ns?
4. What should the depth setting be, in nanoseconds time, for a probing depth of 3 m where the calibrated transmission velocity is 0.095 m/ns?
5. You take over a survey from a colleague who tells you that yesterday's WARR calibration gave a relative permittivity of 25. Your colleague is concerned because the data does not seem to contain as many targets as expected, all of which should be around 2 m in depth. He/she has been using a predetermined depth setting which the technical specification sheet tells you is equivalent to 60 ns. Can you do any better?

SAMPLING ALONG THE LINE OF TRAVEL OF THE GPR

When creating any image, the number of pixels (data points) is critical to the formation of that image and also underlies our ability to recognize the image for what it is. (see, for example, Dennis, 2004, for an illustration of this process). In GPR surveying, there are two critical parameters which determine the number of data points. The first of these is the sampling interval along the line of travel of the radar, sometimes called the trace increment (where the trace is each signal transmitted) or the read distance (where each signal is considered to be one reading). The second critical factor is the space between one line and the next, i.e., the spacing between one survey line (or transect) and the next.

The first consideration in setting the sampling interval is that the physical space between one signal and the next should not be larger than the size of the target. If signals can be transmitted on either side of a target, there is a chance that the target will not be detected. The sampling interval needs to be of the order of or slightly smaller than the size of target to be detected.

This does not mean that the sampling interval should always be the smallest setting available on the radar. Most GPRs are capable of being used with more than a single frequency of antenna so it should not be assumed that every setting on the control panel is optimized for every antenna. While it may be possible to use every setting with any antenna, the parameters which best suit a high-frequency antenna do not necessarily reflect the best settings for a lower frequency one. In addition, GPR controllers vary in the speeds at which they generate the signals they transmit. It may simply not be possible to transmit a wavelength of the size required within the time limit set by the size of the sampling interval. In this case, data will be lost because not all of the signals required will be transmitted as planned. Usually this results in some form of error message but there are still systems in use which do not display an error signal and the problem is only discovered when the length of the survey line in the data is found not to match its physical length.

Even if data are not lost, the other result of using too small a sampling interval will be a slowing down of the radar in order to make sure that the full data set is acquired. This, in itself, is not ideal. A lot of development effort in recent years has gone into speeding up radar performance in order to increase the cost effectiveness of the equipment. It makes sense not to counteract this by using a higher level of sampling than is actually necessary for the task in hand.

It therefore makes sense to take a positive decision about the sampling interval, rather than leaving the radar on its default settings. The major consideration is that the sampling interval should be sufficient for all targets to have a good chance of being on the receiving end of at least one signal. If a lower setting is used and the radar can only collect the full amount of data by moving slowly along the survey line, then it is extremely likely that a higher setting, within the detection constraint, would be more appropriate.

TRANSECT SPACING (SAMPLING ACROSS THE SURVEY AREA)

Transect spacing, or the gap left between adjacent survey lines, is something which varies widely from one application of GPR to another. A great deal depends on whether it is considered necessary to have full area coverage and what can be agreed upon as full area coverage. Even very high-frequency GPRs are blunt instruments because the wavelengths emitted determine what can and cannot be detected. Add to this the change in spread of the beam with depth and, even without considering the electronics lying behind the operation of the radar, it should be evident that there is an inherent uncertainty in the definition of targets.

There is no general agreement on what constitutes acceptable area coverage although there are guidelines available within certain applications (see also Chapter 11: Common GPR Applications). There are still some GPR operators who consider that one or two lines are adequate coverage from which to draw detailed conclusions. Although a great deal depends on the information required from the radar, detailed information is not normally a result of anything other than detailed survey. A useful rule of thumb for close survey is to allow a spacing between successive lines of no more than the width of the antenna being used.

It is worth being aware that the design of all GPRs involves sampling of digital signals for which there is a minimum specification for being able to recover full information, known as the Nyquist theorem. It is not the purpose of this chapter to discuss this but it is important for GPR users to be aware of both this and that data density matters. Widely spaced or single survey lines are potentially of limited use and it is important to assess the risk of nondetection of targets with this in mind. A good example of experimental work within an archeological context illustrating the differences in target imaging can be found by Booth et al. (2008). It is not only in archeological contexts that this issue matters. Within the utility detection industry there is usually considerable pressure on surveyors to complete their work rapidly and as cost effectively as possible and this may lead to minimal use of GPR. Given that the subsurface of all towns and cities is becoming congested with increasing numbers of pipes and services, including plastic pipes which are not readily detectable by other means, it is important to put in place a realistic GPR survey strategy, one which can be discussed with the survey client and that all parties should be aware of the potential for nondetection where this exists.

GRIDS AND GPS

All surveys require reference points so that they can be related to the landscape and environment in which the data were collected. Without this, it would be impossible to locate the target material in the aftermath of the survey. Where more than a single GPR survey line is recorded, the traditional method of GPR surveying is to lay out a reference grid along two orthogonal axes. The size and shape of the grid are determined by that of the area to be surveyed. The spacing of the lines within the grid are determined by the transect spacing between adjacent survey lines (see also previous section). The position of the grid reference lines would typically be recorded relative to permanent reference points in the vicinity so that targets can be found (and/or avoided) afterwards. If a three-dimensional data cube is required in order to map targets, the grid is the basis on which this data cube is formed (see also Chapter 10: Three-Dimensional Data).

GPS can also be used in conjunction with a GPR to provide an absolute frame of reference. This is particularly appropriate for road surveys where the data are collected using a vehicle along predetermined tracks, and also for the low-frequency investigations such as environmental and mining where the surface may not be suitable for systematic survey and where, because of the long wavelengths used, the gaps between survey lines may be of lesser importance. The targets are usually depths of particular material strata which can be easily measured from individual profiles. If the extrapolation to measuring quantities of material is to be made, then there will need to be a denser data set.

It is also possible to use GPS to define a grid. The choice of reference system depends on what is most appropriate for the site and for the GPR equipment being used. Some GPR systems record and store GPS or total station data simultaneously with the GPR data.

ANTENNA POSITION

Since antennas come in finite sized boxes, it is important to be able to relate the position of the antenna to that of the data it produces. The position on the antenna which corresponds to the data being generated is its midpoint, i.e., the central point between the transmit antenna and the receive antenna. For monostatic antennas, it is the midpoint of the single antenna used for transmission and reception. This position is normally marked on the outside of the antenna casing so that the radar can be aligned with a predetermined position on the ground.

Some GPRs are set up to run for a specific distance. For these radars, it is important to align the midpoint of the antenna with the starting line of the survey grid. For other systems, it is not necessary to start on the reference line. In this case, a marker function may be used to mark when the midpoint of the antenna crosses the start and end lines of the survey area.

Where all survey lines begin along the same reference line, the GPR data will automatically align if two or more radar traces are to be combined into three-dimensional data. For data where a marker has been used to indicate the position of the reference line, the processing software will allow the data to be aligned along the markers on the data, i.e., the position corresponding to either the start or finish lines of the survey area.

Most GPR systems have software which allows the operator to collect and in some cases, process into a three-dimensional data block, the data from a grid-based survey area.

SETTING PRIORITIES

A practical way of dealing with the balance between full data collection and the cost effectiveness of the survey is to set priorities for different areas of the site. If GPR is the only geophysical tool to be used on site, then ideally, it should be used to maximum effect but this does not preclude allocating resources efficiently. A typical example of setting priorities can be illustrated from the GPR survey of a main road in order to provide the necessary information about the current state of the road structure and its underlying subsurface, including any potential problems developing. It is not unusual to use the GPR to survey the inside wheel track, the center of the lane and the outer wheel track for each traffic lane. Although this does not constitute total coverage, it does demonstrate a strategic judgment as to the level of information necessary.

Similarly, in utility detection, it is possible to base a utility survey on the information gleaned by observation and enquiry, supplemented with the use of electromagnetic methods and then form a judgment on the extent of the GPR coverage.

An archeological example of setting priorities can be seen in the survey of a Peat Moss at North Ballachulish in the Highlands of Scotland (Clarke et al., 1999). The whole site was surveyed using relatively widely spaced survey transects to establish whether or not there were possible archeological remains in any sector. A closer spaced survey was then centered on the area in which anomalous signal returns were observed. Had the distribution of archeological remains been more widespread, it is likely that the strategy would have had to be adapted accordingly.

In all cases, the risks and results of nondetection should form a part of the strategic plan and it is important that the client is as aware of the strategy being applied as the GPR surveyor.

INFORMATION REQUIRED PRIOR TO SURVEY AND DECISIONS TO BE TAKEN

The various points discussed above illustrate the range of information which it would be useful to have in advance of survey. In the real world this is seldom, if ever, fully available. It is useful to have a pro forma checklist which can be used on any survey. This allows the surveyor to decide on what GPR and other equipment is required, whether additional antennas should be brought in reserve, to highlight what the inherent risks of an unsuccessful survey might be and ultimately contributes to the costing as well as the planning of the survey.

The checklist should cover the following:

- Nature of target—i.e., its suitability for detection within its environment
- Target size

- Probing depth
- Soil conditions
- Groundwater
- Surface conditions
- Survey direction
- Sampling parameters including transect spacing
- Site access

As discussed above, the primary issues include the depth, size, and type of target(s) since upon this depends both whether detection by GPR is likely to be successful and also the frequency of antenna to be used. It is useful to know something about the nature of the soil or other material which lies above and around the targets. This is necessary so that we can be sure that the target will be detectable but also has implications for selecting the optimal frequency of the antenna(s). For example, of particular concern is the presence of reinforced concrete because of the potential conflict between having a short enough wavelength to penetrate between metallic reinforcement bars and one which is long enough to reach the full depth of the underlying targets. Also, if there is a significant proportion of clay in the soil composition, what is the likelihood of water content and is there a risk that the soil, if wet, may be unsuited to GPR? Will a lower frequency/longer wavelength antenna be required because of the soil conditions? Unfortunately, although borehole information may be available, this is not always expressed in a manner which is easy to translate into potential radar losses. For example, it is not unknown for a borehole log to record a depth of "clayey sand" followed by a depth of "sandy clay" of one color followed by another depth of "sandy clay" of a different color. The information required by the GPR surveyor is the electromagnetic response of the soil, preferably under both dry and wet conditions. Difficult as it is to extract this from borehole information, it is even more common to have no soil information at all.

It is useful to know the depth of groundwater relative to the potential targets. This allows the question of probing depth to be considered in relation to a possible decrease in transmission velocity. The question of potential signal losses can only be considered if the nature of the soil is also known or soil samples have been taken.

The condition of the surface is important whether the survey is to take place indoors or outside. Maintaining ground coupling is an important factor in any GPR survey and planning a grid around surface obstacles which could be anything from the base of an overhead walkway to trees or, in an indoor context, furniture fixed to the floor, might have an impact on the transect spacing, for example. It is usually possible to work around small numbers of isolated surface obstacles. Site clearance in advance of the survey may be necessary if the obstacles are many in number or if leaving them in place would result either in an inappropriately low level of data collection or the antennas being positioned randomly relative to the ground as they move across it. If the antenna angle changes continually relative to the ground surface, the area of investigation will not be the actual subsurface and there will be breaks generated in the continuity of the data, potentially resulting in misleading information. There is also the question of how the survey is to be carried out whether on foot or from a vehicle. In the case of unusually rough ground, it may be necessary to customize the antenna both to maintain ground coupling as much as possible and to prevent damage to the antenna during the survey.

Survey direction is a function of the type of GPR application and the orientation, if any, of the targets. It may be possible to survey in one single direction or it may be more appropriate to carry out the survey in two orthogonal directions. The decision should be taken in advance and discussed with whoever is commissioning the survey.

Not all GPR surveys are area surveys but, for those that are, the spacing between consecutive lines should be decided in advance of survey if possible. The deciding factors are the size and nature of the targets and the requirement, if any, for three-dimensional data. This may have to be modified as the survey progresses, depending on the results obtained but, as this forms part of the survey sampling strategy, it makes sense to have a coherent plan from the start.

Site access to the survey area is also an important factor to establish in advance. This varies widely depending on the application. For example, the question of whether road closure is to be arranged for the duration of the survey applies to road and utility surveys but is extremely unlikely to apply to a mining resources investigation. For the lower frequency investigations including archeology and environment assessment, the question may simply be how to reach the survey site and whether vehicles can be used to do so. For large surveys it will also be necessary to find suitably secure areas for storage of vehicles and equipment.

THE UNSUITABLE SOIL RISK

As mentioned previously, it is not always possible to gain all the information necessary in advance of a survey and the information which is most commonly missing is the electromagnetic characteristics of the soil. A suitable strategy for dealing with this possibility is to put a rider into any quotation to the effect that there is uncertainty regarding the suitability of the soil. This rider should state that the GPR surveyor will do their best to obtain meaningful results but, should this prove impossible, they will withdraw from the site, having first informed their client of the observed unsuitability of the soil. Under these circumstances the surveyor then limits their charges to those for mobilization to and from the site, for the actual time spent on site and any processing time, agreed with the client once the difficulty has become apparent. Processing time might be necessary to confirm the unsuitability of the soil.

In this context, "doing one's best" might include allowing for a lower frequency antenna to be substituted, depending on the site conditions.

ANSWERS TO THE EXERCISE

1. Speed = distance/time = $(2 \times 1 \text{ m})/20 \text{ ns} = 0.1$ m/ns.
2. Speed = distance/time so time = distance/speed = $(2 \times 2)/0.08 = 4/0.08$ or 50 ns. Remember that distance = two-way travel time.
3. As for the previous question, time = distance/speed = $(2 \times 2)/0.15 = 4/0.15$ or 26.67 ns.
4. Once more, time = distance/speed = $(2 \times 3)/0.095 = 6/0.095$ or 63.2 ns.
5. Obviously you will calibrate velocity at some point today but, assuming there are no great changes in the weather and that either your site is similar in soil conditions to your colleague's or you are about to repeat his/her survey, you can use yesterday's calibration for planning purposes. $V = c/\sqrt{\text{Er}} = 0.3/5$ or 0.06 m/ns. What does this say about your site (cf Chapter 4: The Effect of Water and Air)? The time setting used has been 60 ns. Speed = distance/time so the distance $(2 \times \text{depth probed}) =$ speed $\times$ time $= 0.06 \times 60$ or 3.6 m. The depth probed is therefore only 1.8 m and it is not a great surprise that the targets are not being detected.

REFERENCES

Booth, A.D., Linford, N.T., Clark, R.A., Murray, T., 2008. 3-dimensional multi-offset ground-penetrating radar imaging of archaeological targets. Archaeol. Prospect. 15 (2), 93−112.

Clarke, C.M., Utsi, E., Utsi, V., 1999. Ground penetrating radar investigations at North Ballachulish Moss, Highland, Scotland. Archaeol. Prospect. 6, 107−121.

Dennis, T.J., 2004. Imaging geophysical data—taking the viewer into account. Archaeol. Prospect. 11, 35−48.

Chapter 8

Data Processing

Chapter Outline

It is not the purpose of this book to describe all possible data processing steps. This is something which has to be learned by experience within the context of one's own survey environments and equipment. The techniques which are most appropriate vary with the GPR equipment used and also from one site to another. Individual GPR surveyors also frequently have their own personal favorite methods of presenting the data. Some of the more commonly applied techniques are discussed in this chapter in order to illustrate the sort of simple techniques which can be applied in order to highlight targets. As a processing sequence, they form a good place to start. If the results are not as good as required, then it is time to start investigating alternatives and building up a library of further appropriate techniques. The only really important rule to observe is to present as clearly as possible with the minimum amount of processing appropriate to the data set. For some data, this will mean very little processing but, in more challenging cases, it may mean the application of considerably more.

Some GPR systems apply processing filters directly to the data as it is collected. This is perfectly acceptable practice but it is essential that the operator understands which processes are being applied and selects a suitable processing package. Use of the wrong filter set can effectively eliminate targets. It is important to understand how the data are affected by the prefilter set and not resort to using the default setting for all surveys. Fig. 8.1 illustrates the problem. The same line of data has been processed twice using exactly the same processes with the exception of the type of background removal applied. For the top trace, a standard background removal has been applied. For the lower trace, a subtracting average process has been used as background removal. The result of using an inappropriate form of

Ground Penetrating Radar. DOI: http://dx.doi.org/10.1016/B978-0-08-102216-0.00008-4

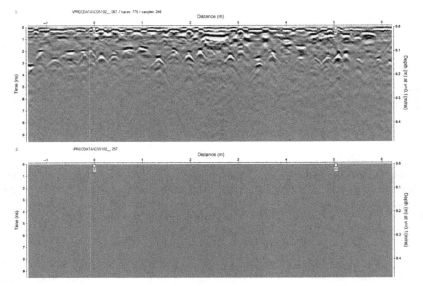

FIGURE 8.1 GPR data recorded across a mosaic using a standard background removal (above) and subtracting average (below).

background removal is to take away all evidence of the mosaic which was surveyed. The upper radargram clearly shows the depths of the major elements in the mosaic. The lower radargram is useless. It is important to understand the effects of any directly applied filtering before using the radar.

SENDER/RECEIVER CORRECTION

Once the collected data have been imported into the preferred processing package, the first process to consider is whether a correction is required for shallow depths to allow for the geometry of the transmitter, the receiver, and the target. For bistatic antennas, there is a physical separation between the transmit and receive antennas. The theoretical travel path of the signal from the transmitter to the target and back to the receiver is illustrated in Fig. 9.1. This demonstrates that the real depth of the target is not the same as the two-way travel time (transmitter to target, target back to receiver) but is, in fact, slightly less. There needs to be a geometrical correction to the data, based on the distance between the transmitter and the receiver and making use of the transmission velocity on site (see also Chapter 9: Antennas and Screening).

The difference between the two-way travel time and twice the real depth becomes vanishingly small as the depth of the targets increases. The effect is greatest in the near subsurface. Where a monostatic antenna is being used, there is no need for an adjustment since the same antenna serves as transmitter and receiver.

There is no need for complicated calculations since all GPR processing packages make provision for this correction. The size of the gap between the center of the transmit antenna and that of the receiver antenna is usually stated in the technical specifications. Some GPR systems store this data in the operating software also. The transmission velocity should ideally be the calibrated velocity but an estimate may be acceptable for this purpose.

TIME ZERO REMOVAL

All GPR data must have Tzero (Tz) removed before any further processing is done. This is because above the direct signal, the transmission is internal to the radar and not related to the subsurface. If Tz were to be left in the depth measurements, whether in nanoseconds or centimeters/meters or feet/inches, then the depth measurements would be inaccurate, overstated by the value of Tz. On some GPR systems this can be done by adjustment within the operating software, on others it may be automatically adjusted on import into the processing software. In all cases, it is possible to use the processing software to remove any residual value.

Although Tz can be assumed to relate to the uniform monochrome "layer" above the direct signal, it is best measured by examining the individual traces (or A-scans) to find the first zero crossing of the signal. An example of this is shown in Fig. 8.2. The cursor is drawn down through Tzero on the data until it reaches the interface with the direct signal, in this case indicated by the top white band. At the same time, the A-scan or wiggle trace is used to identify the point at which the signal crosses the central axis. In this case, the signal crosses from left to right across the axis. Tz can be read off the software. For example in Fig. 8.2, the reading is 2.59 ns.

Something else to be aware of is that a good GPR system generally has a stable Tz, not one that fluctuates. If there is evidence of fluctuations in Tz, then it is possible that either the antenna or another part of the system needs repaired. An unstable Tz would affect the reliability of the depth readings.

BACKGROUND REMOVAL

By definition, all GPR targets, whether ducts in the ground, mineral ore deposits, graves, or unexploded ordnance, are anomalies within the context of their environment and the processing steps applied are aimed at highlighting the anomalous material at the expense of the more general environment. Background removal is a method of removing the elements of the common environment from the data in order to leave any anomalous material more easily seen.

There are many different methods of background removal. The background to be removed can be defined across the full radargram or from one part of the radargram. There are usually options to average over a number of

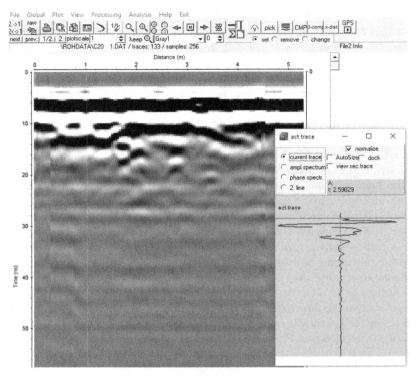

FIGURE 8.2   Measuring Tz using an individual wiggle trace.

traces in a variety of ways. It is usually best to start with a simple version and only use more complicated methods if they are actually required. Fig. 8.1 shows an extreme example of what can go wrong if an inappropriate method of background removal is applied.

One thing to be wary of is using background removal based on the full radargram if there is an area of very strong signal within that radargram, say from a metal object or an air gap. The result of doing this tends to be a spread of the stronger signals across a wider area than was originally affected. As this could potentially distort the interpretation of the data, it is wise to select only a part of the radargram for the definition of the common background where there are one or more discrete very strong signals in a data set. This may mean checking the full data set in order to be sure there is a common area for each line of data which does not contain high amplitude signals. Once a suitable area has been selected, this is used to define the background common to all of the data to be processed. This background can then be removed in the normal way.

The effect of background removal is to highlight any anomalous material within the data set. Visually it also normally breaks down the strong banding from the direct signal. Although this is usually very useful, it may not be helpful if the target of the survey is a near continuous layer. The continuity of the layer could result in its removal. Discrete objects, on the other hand, will remain visible. If layer information is the purpose of the survey, background removal may not be a suitable processing step to apply.

GAIN

As each signal passes into the ground, some parts are lost into the environment (attenuated) and others reflected back to the receiver so that, with increasing depth each signal diminishes. If no compensating gain is added to the data, any targets in the lower part of the depth probed by the radar will either not be visible or only faintly so. The gain which is applied to the data is usually referred to as "time-based" gain in recognition of the process of the loss. There are many different methods of compensating for lost signal. The method which suits best will depend on the electromagnetic response of the ground (or other material surveyed), on the probing depth of survey and, to some extent, on the GPR system used.

The two major types of basic gain are linear and exponential. Linear gain applies a constant level of gain to the full signal throughout the full depth. Exponential gain applies increasing amounts of gain with increasing depth. Generally speaking attenuation is exponential with depth and it therefore makes sense to begin with the application of exponential gain. Linear gain is sometimes more appropriate in low-frequency investigations. There is no single solution and the GPR surveyor has to experiment to find out which system and which value of gain best suits his/her data set.

The aim when applying gain to a data set is to make all targets as visible as possible without increasing the inherent noise within the data set. The best method of doing this is to observe the effect of the gain on a trace (or A-scan), increasing the gain until the target area of the signal is made more visible but stopping before the very lowest part of the signal begins to increase to an unrealistic level. Fig. 8.3 shows the effect of changing the gain on one individual radar trace. In each of the three images, the first long

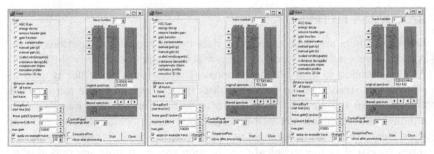

FIGURE 8.3   The addition of exponential gain, from left to right: too little, too much, enough.

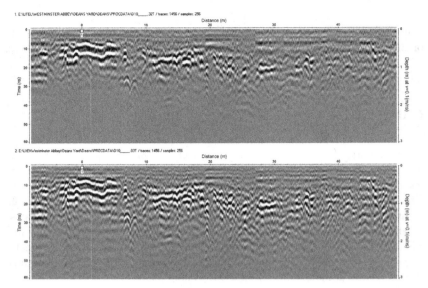

FIGURE 8.4 Radargram without the addition of gain (above) and with an appropriate amount of gain added (below).

blue box shows the original signal and the second the effect of adding gain. In the left-hand image, too little gain has been added and the second signal is almost identical to the first. In the second image, too much gain has been added. The level of the signal in the target area has increased dramatically but so too has the lower end of the trace. The application of this amount of gain would result in very messy data. In the third, right-hand, gain window, enough gain has been added to enhance the target area of the signal without making a dramatic difference to the deepest part of the signal. This is what we are aiming for.

The result of adding gain is shown in Fig. 8.4. Although the original data are reasonably clear, the addition of a small amount of gain brings up a few targets lower down in the subsurface.

BANDPASS FILTERS

Bandpass filters are used to clean up data by the removal of spurious noise. Again, there are many types of filter which all operate slightly differently. An easy version to apply is the Bandpass Butterworth. This filter is used to remove any signals whose frequency does not correspond to the range of the GPR used in the survey. There is no agreed definition of a radar's range although all manufacturers will describe their equipment, correctly, as ultra wide band. The frequency range, as opposed to the peak frequency, is sometimes specified in the technical specification sheets. If not, one option is to define the range of the antenna as $0.5-2F$, where F is the peak frequency (i.e., the frequency associated with peak energy). So, for a 500 MHz antenna, the range would be 250 MHz to 1 GHz.

It is not always necessary to use a bandpass filter but it can sometimes be helpful in cleaning up data, particularly where large amounts of gain have been added, typically where the survey took placed over lossy or uneven ground.

STACKING

Stacking traces is another process which is not always necessary but which can sometimes be useful. The basic aim of stacking is an averaging process over a number of traces, intended to make targets clearer and to reduce any associated spurious noise. Some manufacturers offer this as an option within the operating software and at least one manufacturer's system handles stacking internally within the antennas during data collection. In neither case is it necessary to use stacking in the processing. The latter approach preserves the operating speed of the radar, the former can lower the operating speed. For all radar systems, stacking of traces can be applied in postprocessing.

MIGRATION

As discussed in Chapter 5, Velocity Calibration, migration is used to mathematically collapse hyperbolas into single points. This is particularly useful where linear objects are the target of the investigation, e.g., pipes and services. It defines more accurately the position of this type of target without the potential spread caused by the hyperbola (which is caused by the beam spread of the radar) (cf Fig. 5.2). It is not, however, necessary for all survey data and can cause unwanted effects in inappropriate circumstances.

As with the other processes mentioned in this chapter, there are many different migration techniques. The one described here is migration by diffraction stack which is an easy example to apply. For this method there are two defining factors, the summation width of the hyperbola and the transmission velocity. Curve fitting is used to define the summation width by altering this value until a curve is obtained which fits not only the outline of the hyperbolas in the data but also their visible length. The easiest way to do this is to start with a default setting, say 50, and alter it until a good fit is obtained. Applying the migration to the data will then demonstrate whether or not the transmission velocity requires adjustment, effectively refining the calibration. As seen in Chapter 5, Velocity Calibration, if the hyperbola collapses to a neat point, then the correct velocity has been applied. If there is a slight upward curl to the point, then the velocity requires to be reduced and if there is a slight downward curl, then the velocity should be increased slightly.

SEQUENCE PROCESSING

There is no need to process every line of radar data individually. All processing packages allow the GPR surveyor to process full data sets at the same time. One file should be selected from the data set, preferably a file that is reasonably representative. This file is then processed experimentally until the GPR surveyor is happy that the presentation of the results is good enough for the task of interpretation to begin.

The processing sequence can be recorded and stored as the processes are applied. Any that are not deemed satisfactory can then be removed in the same way until finally a processing sequence for the full data set can be saved. The remaining files can then be processed as a batch, making this a much faster exercise. As explained below, if there were above ground obstacles impeding the survey, it is usually best to process the data before repositioning the files collected around the obstacles relative to their actual positions.

It is useful to be aware that data can be distorted inappropriately by postprocessing. In case of doubt, the easiest way of being sure that this is not the case is to compare the final processed data with the features visible in the original raw data.

OTHER USEFUL TECHNIQUES (X-FLIP, CUT AND PASTE)

Most GPRs will allow data collection in any direction although at least one system is set up for all data to be collected along lines lying in the same direction. When the survey area is small, then it does not matter whether the GPR surveyor returns to the original baseline for each successive line of data. However, for large survey areas, this can represent a waste of time and it is generally easier to collect data in one direction and then record the next line in the opposite direction. The process of collecting data in alternating directions is sometimes referred to as "meandering." It is then necessary to change the direction of at least half of the data set so that the lines lie in a consistent direction. *X-Flip* allows the GPR surveyor to change the direction in which the radargram is displayed so that all of the survey lines run in the same direction. This is essential if a three-dimensional (3D) data cube is to be constructed (see also Chapter 10: Three-Dimensional Data). It is often also possible to change the direction of the data on import into the processing package although this can result in all of the data being displayed in the opposite direction to that required for 3D processing (see Chapter 10: Three-Dimensional Data).

Cut and Paste is a process used to separate and recombine files in their respective positions to allow for working around large obstacles within the survey area. The principle is the same whether the obstacle is a tree in a field, the base of an overhead walkway where it reaches the pavement, a pillar inside a church, or a temporary enclosure erected for the overnight security of equipment in the center of a development site. The GPR data are collected along lines either using a grid or GPS, whichever is the most appropriate. On reaching the obstacle, a marker is placed on the data. The GPR surveyor then moves around to the other side of the obstacle and continues the survey line, marking the data with the start position. During this process the last position of the radar before moving round and the first position of the radar after moving round must be marked on the ground so that this gap can be measured. The position to be marked on the data and on the ground in each case is the midpoint of the antenna casing in each of the two locations. The distance between these two points must be recorded so that the data recorded before the obstacle can be attached to the data recorded after the obstacle with the correct distance in between.

If, as is often the case, the obstacle is large enough to affect several survey lines, then the process is repeated on the next and any subsequent lines which are impeded by the obstacle. In each case the gap between the end position before the obstacle and the start position after the obstacle must be recorded.

When it comes to processing, each file affected by the obstacle is extracted into two separate files on the basis of the markers on the data. The first file runs from the start of the survey line to the first marker. The second file runs from the second marker to the end of the survey line. The middle part, recorded around the obstacle, is discarded. These two files are then recombined by adjusting the start distance of the second file to the correct distance from the end of the first file, as indicated by the first marker. This is the same distance measured on the ground between the stop and start locations before and after the obstacle.

An alternative to using markers on the data is to treat each part of the survey line as a separately recorded file. In this case there is no need to extract separate files for recombining. The ground distance between the files will still have to be measured. The files from each survey line can then be joined together in the same manner as for the extracted files with the distance between the end of one file and the beginning of that on the other side of the obstacle being the same distance as measured on the ground.

Where there are several obstacles, the same process is repeated for each obstacle. Although this can add to the processing time, it does speed up the survey time on site which is more often the critical factor. One thing to be careful about is that processing such as background removal should be done before the paste part of the exercise is attempted. If background removal is applied directly to a combined file with a blank in the position of the obstacle, this will spread the signal across the blank area. If the processed files are recombined, the space will be left blank. Not only does this make it much easier to fit plan views of the GPR data, e.g., to a digital drawing, it also ensures that it is clear that there was an obstacle at this point. Note that the size of the blank area is equivalent to the obstacle itself plus the offset of the radar on either side so it will be larger than the obstacle.

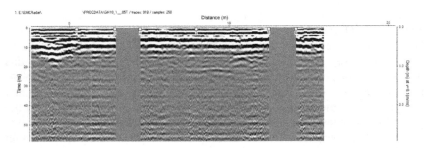

FIGURE 8.5 Example of 2D data with gaps indicating the presence of trees.

Fig. 8.5 shows an example of 2D data which has been collected over open ground where trees are growing. Markers have been used to indicate start and end positions for the radar as it reaches each tree. The recombined data take account of the gaps in the data around each tree.

Fig. 10.4 shows an example of a time slice, extracted from a 3D data block where the data have been collected in and around a development site on which there is a fenced enclosure. The gaps indicate the position of the fence and the offset of the radar trolley from the fence at the start and end positions. It is also obvious where the enclosure entrance is located. Where the gaps relate to permanent features in the landscape, the gaps can be very useful in positioning the time slice accurately into a site drawing.

Chapter 9

Antennas and Screening

Chapter Outline

There are many different antenna designs. For the basic use of a GPR, it is not necessary to know or fully understand the factors which influence antenna design. There are, however, a few basic principles which it is useful to know and this chapter addresses some of these. One thing which is very useful to understand is that the internal workings of any antenna, including both the internal antenna and the screening have to work together so as to define the electron path. In simple terms this means that trying to customize any part of the antenna, apart from external mechanical support, is something that should only be attempted by the manufacturer. Adding materials to act as a screen, for example, will not improve matters. It will alter the delicate balance that dictates the electron path with unfortunate consequences for the quality of the data.

MONOSTATIC/BISTATIC ANTENNAS

Every GPR has at least one transmit antenna and one receive antenna. If the transmitter and the receiver are the same physical piece of equipment, then the system is said to be monostatic. A monostatic antenna may typically be relatively small because only a single antenna is used for transmitting and receiving. Since the transmission takes place from the same location as the returning signal is received, no correction is needed to the data for a gap between the transmitter and the receiver. On the other hand, because there is a small, finite time difference required to change the transmitter into a receiver and vice versa, monostatic antennas have a small built-in time delay which means that monostatic systems operate more slowly than bistatic ones.

Ground Penetrating Radar. DOI: http://dx.doi.org/10.1016/B978-0-08-102216-0.00009-6

Bistatic antennas consist of a separate transmit antenna and receive antenna. This may not be apparent when looking at the system since they are usually housed in a single antenna casing. They do not have the built-in time delay characteristic of monostatic antennas since there is no requirement to switch from transmitting to receiving and vice versa. However, since they consist of two antennas rather than one, they tend to be larger than monostatic systems. The size of any antenna, regardless of the type of system, is directly related to the wavelengths that it will transmit and receive.

Since there is a measureable distance between the transmitter and the receiver, even though they are normally placed side by side inside the antenna housing, an adjustment has to be made to the apparent depth of the data as this is distorted by the geometry of the radar's antennas relative to the target. Fig. 9.1 illustrates this.

The depth recorded by the radar is the two-way travel path from transmitter (Tx) to the target and back from the target to the receiver (Rx). As can be seen from the diagram, the real depth is less than half of the travel path so the radar data is overstating the real depth of the target. The error has a greater effect on shallow depths. The deeper we probe, the closer the two-way travel path comes to being twice the real depth. The adjustment which has to be made therefore depends on the geometry of the radar system. The details of the geometrical calculations needed to correct this do not matter since all GPR software packages already contain the mathematics required to do this automatically. It is sufficient to enter the distance between the center of the transmitter and the center of the receiver and the transmission velocity into the appropriate function and the processed data will be automatically corrected for depth. This correction is referred to as a sender/receiver correction. The sender/receiver correction should be carried out before beginning processing (Chapter 8: Data Processing).

The transmission velocity is essential for the calculation because the radar measures extremely accurately in nanoseconds time and this has to be

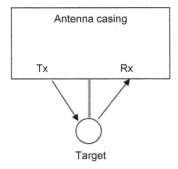

FIGURE 9.1 The two-way travel path (outer arrowed lines). This is greater than twice the depth of the target (central line).

translated into basic measurement units of centimeters/meters or feet and inches. Ideally the transmission velocity used should be the calibrated velocity for accuracy's sake but a preliminary adjustment, based on expected velocity, can be made.

Since increasing the speed of data collection has been of growing important to modern GPR systems, most of the currently available GPR antennas are bistatic although there are still some monostatic systems in use. The distance between transmitter and receiver is sometimes recorded directly in the operating software but, if not, can usually be read from the technical specifications. Failing this, it is possible to estimate the distance by taking a reading with a tape measure. The central point of an antenna is usually clearly marked so that the position on the ground can be related to the data generated by the system. Divide the box into two along this central line. The transmitter is in one-half of the box and the receiver in the other. It does not matter which antenna is in which half. The measurement to take is from the center of one-half of the antenna box to the center of the other side.

BOWTIES, HORNS, AND SNAKES

Transmission of a clear signal out of the radar into another material is a complicated process particularly if side effects, typically ringing, are to be reduced to a minimum. It is not the function of this book to address any of the design issues but it is useful to understand that there are many different shapes of antenna. The three most commonly used in GPR design are the bowties, horns, and snakes. Of these, the most widely used are probably the bowties.

Bowties are so called because their shape mimics that of a bowtie. In reality, the actual shape of the antenna varies from one manufacturer to another and, taken together with the screening (see below), is both integral to the transmission method and also part of the manufacturer's protected design. The central crossover point of the bowtie can be considered as the transmission/reception point.

Most ground-coupled systems use bowtie antennas, irrespective of the frequency. The flattened shape is intended for ground-coupled operation. The bowtie sits on the under surface of the antenna. Any screening will be placed around and above (Fig. 9.2).

Horn antennas come in many shapes and sizes and are generally used for transmission into air. This means that they are frequently used, for example, on road investigations where the survey is carried out by a vehicle, at speed. Although they can be used for lower frequencies, they are most frequently used for frequencies of 1 GHz and above.

Fig. 9.3A shows a simple example of a horn antenna, within its screen. Fig. 9.3B shows another popular variant of a horn antenna, known as a Vivaldi antenna. In both cases, the transmission point is from the inside at the top of the antenna. Like the bowtie antennas, these antennas are usually

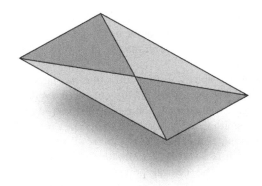

FIGURE 9.2 A basic bowtie antenna.

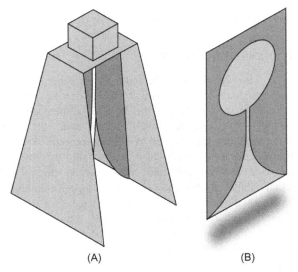

(A)   (B)

FIGURE 9.3 (A) An example of a horn antenna. (B) An example of a Vivaldi horn antenna.

screened (see also below). Whereas the bowtie antenna shape is intended to maintain ground coupling, it is evident from the shape of the horn antenna that transmission is directly into the air.

Snake antennas are used for the opposite end of the GPR spectrum, i.e., for very low-frequency investigations. As their name suggests, the antennas are long and thin. They are generally towed behind the GPR surveyor. This makes them suitable for tackling uneven terrain which is often also a feature of very low-frequency GPR applications. The reason for the shape of these antennas is the lengths of wavelengths that they transmit. All radar antennas are related in size to the wavelengths that they transmit. For the very low

FIGURE 9.4   Snake antenna in use on a Chinese hillside.

frequencies, producing a bowtie to the equivalent dimensions of the wavelengths would make the antennas very unwieldy, difficult to transport and cumbersome to use. It would also be virtually impossible to draw the antennas over the landscape in such a way as to maintain efficient ground coupling. The snake formation is much lighter and easier to tow but can also mold into the formation of a varied landscape, and therefore maintain reasonable ground coupling. Their principal drawback is that, because of their size, they are usually used unscreened although, as discussed below, this is often not critical in the environments in which they are used (Fig. 9.4).

BOREHOLE ANTENNAS

There is more than one type of borehole antenna. Most borehole antennas are designed to be lowered into a borehole in the ground in order to detect from a lower level than the ground surface. This is particularly useful where the soils overlying the area of interest are not entirely suited to investigation by radar since the borehole effectively bypasses the overlying layers. It can also be useful if target definition requires a higher frequency than would be possible if investigation were to be carried out from the ground surface since the distance/depth to target is less. They are most commonly used in mining applications to detect mineral seams or subsurface fractures but have also been used for examining the integrity and position of sheet metal piles and other deep foundations. They are not as commonly in use as the other types of antenna described above but are readily available. The frequency ranges vary with the tasks for which they are used but are typically somewhere in the mid-frequency ranges of 100 MHz up to 900 MHz.

An alternative design for another specialist use is an antenna which can be inserted into a duct or pipe for use either in pipe inspection or to examine the environment around the pipe or duct for the build-up of voids, usually caused by localized water leakage. The shaping of the antenna is dictated by the diameter and shape of the pipe and these are usually high-frequency antennas both for the target definition needed (small targets) and for the limited probing depth required. The survey would usually be preceded by sweeping the interior of the pipe or duct with a mechanical object similar in shape and size to the antenna, known as a "pig" in order to clear any debris out of the path of the radar. This is important to avoid damage to the antenna and to avoid it getting stuck as well as any impact on the survey data.

Ground coupling of the antenna is just as important for this application as any other GPR survey in order to avoid surface effects generated within the pipe or duct from the movement of the radar. This can be achieved by spring loading of the casing. It is also usual to include a camera in the unit so that the operator can see what is happening in the subsurface. Control mechanisms are usually left outside of the pipe or duct. The antenna may be hand drawn through the pipe or duct or it may be towed if a suitable crawler system is available. The other important factor in this type of survey is designing the equipment to be explosion proof. Although this is not always necessary, it is a factor to be taken into account, particularly in sewers where there is methane present.

ANTENNA ARRAYS

Not all bistatic systems use a single transmit antenna and a single receive antenna. Antennas which are combined in phase with one another can act to narrow the beam of signals transmitted and reflected. A number of manufacturers use dual arrayed antennas for this purpose. Although there is a small reduction in beam width and corresponding improvement in the clarity of the data, it is not possible to reduce the beam to a single line by adding multiple antennas. This would, in any case, increase the size of the antenna casing which would not be helpful for ease of practical use.

SIZE MATTERS—THE MANUFACTURER'S RULE OF THUMB

One thing which should have become apparent from the description of the various antenna types is that the size of the antenna is directly related to the wavelengths (and hence to the frequencies) that it is designed to transmit or receive. It is useful to be aware of this when assessing the likely performance of a GPR antenna. Where overenthusiastic performance claims are made, it is therefore very useful to know the manufacturer's rule of thumb that the width of an antenna is approximately half a wavelength in size. Add to this the knowledge that, except in certain environments, there is very little chance of gaining more than 20 wavelengths in probing depth and any potential user or purchaser of a radar system ought to be able to work out for himself/herself what the real capability of any antenna is. As a result, they should also be able to avoid overexaggerated performance claims in favor of more realistic assessments.

RADIATION PATTERNS

So far transmission and reception of signals have been presented as though it is natural for any radar, equipped with a suitable antenna, to radiate into the ground. This is not the case. The normal radiation pattern of any antenna is through 360 degrees (Fig. 9.5).

In order to obtain a beam which is directed into the ground, antennas require screening. Without screening (sometimes referred to as shielding), the antenna would radiate as much above ground as below the surface. The data generated as a result would include any movement of the GPR operator and all aboveground obstacles as well as the anomalous material below ground which was the object of the survey. The resulting images would be potentially extremely confusing and difficult to interpret.

SCREENING

The size and shape of antenna screens varies considerably from one manufacturer to another and is an integral part of their protected intellectual property rights (IPR). The simplest screen shape forms a box sitting on and connected

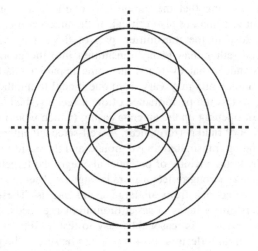

FIGURE 9.5 The radiation pattern of an antenna.

to the antenna blades. In this way the screen forms five sides of the antenna with the antenna blades forming the sixth (lower) surface. In the case of a bow-tie antenna, the sixth surface is the one in contact with the ground. For a horn antenna, the horn shape is normally suspended within the five screen sides so that transmission is made through the sixth, the lowest antenna surface.

The screen or shield must surround all of the antenna except for the working surface which will be placed in contact with the ground (or other surface being investigated). The size of the screen and its distance from the antenna blades are dictated by the wavelengths transmitted or received by the antenna. The size of the screen or shield is therefore directly related to wavelength also. In order to block out unwanted transmission and reception, the screens are made of metal since radio waves do not penetrate metal. Neither the antenna blades nor the screen are visible from the exterior as both are surrounded by an external antenna casing. This should never be removed except by a designated competent electronics engineer or technician. Apart from the legal questions of IPR ownership, vested in the manufacturer, opening up the antenna casing risks damaging the balance of the internal electronics of the antenna and screen.

It is important to understand that antenna screens are an integral part of the transmission system. Taking an inadequately screened antenna and adding a metallic surround to it would simply change the electrical path inherent to the system. This is potentially disastrous in terms of the data then produced. Not only will the data be messy and difficult to read, since it is impossible to determine the path of the signal, there is also a major risk of generating false targets. The only solution to an inadequately screened antenna is to acquire one with better shielding.

Although the point is more critical for a GPR designer than a user, it is probably worth knowing that the dimensions of the shield and its distance from the transmit antenna are also critical. If the antenna box is too low and the screen too close to the transmitter, part of the signal will be diverted toward the screen rather than being transmitted into the ground. All a user needs to understand is that even for the same apparent characteristics, the performance of an antenna may vary from one model to another.

It is possible to test the performance of the screening/shielding without the drastic measure of opening up the antenna casing (which would not in any case give any indication of field performance). A well-screened antenna will prevent primary signals being transmitted above ground and the easiest way to test this is to use the GPR in the vicinity of potential aboveground targets. Metal, being a conductor, is a very strong reflector of radar signals. There is therefore a risk of aboveground reflections from all large metal targets. These may include fences, parked cars, many items of street furniture, large sheds, or even aircraft not currently in service. It is reasonably easy to test a GPR in the vicinity of some of these potential reflectors. If the radar can be seen to be picking up signals reflected from any of these aboveground surfaces, then the screening could probably be improved. As the GPR operator approaches the aboveground target, he/she may be able to see an on-screen reflection which apparently becomes shallower, the closer the radar gets to the metal. Moving further away, the same signal will indicate an apparent increase in depth as he/she retreats from the aboveground target. In reality the apparent "depths" are the distance between the radar and the obstacle. It is the consistency of the signal with the radar's position relative to the potential reflector that reveals that the screening is not as good as it could be. Most mainstream commercial GPRs are well screened and should pass this test although the quality of the shielding does vary from model to model. Even working with a well-screened antenna, poor soil conditions can aggravate the likelihood of aboveground reflections. Potential purchasers of GPR kit should also be aware that cheaper antennas may not have as good screening as those of the mainstream manufacturers.

If the target is overhead, as would be the case for internal metal beams or aircraft wings, there will be a constant reflection at an apparent depth. The depth will equate to the height of the beams or wings above the radar, translated at a transmission speed of 0.3 m/ns since the material between these potential reflectors and the radar is air and not soil. Similar tests can be carried on nonmetallic objects such as walls but the inherently strong signal responses of potential metallic targets make for easier perception.

Note that working in an indoor environment it may not be possible to avoid secondary aboveground reflections even from a well-screened antenna. This occurs when signals pass into the ground and, because of either the electromagnetic response of the ground or that of strongly reflective near surface buried objects, some of the electromagnetic pulses are transmitted into the air and re-reflected from the roof of a building (especially one with overhead metal beams). This type of signal can usually not be eliminated and the GPR surveyor simply has to be aware of the possibility so as to avoid misinterpretation.

COMMON SOURCES OF EXTERNAL INTERFERENCE

There are a number of aboveground targets which it is difficult, if not impossible, to eliminate and which will interfere with the GPR data. Overhead power lines are a common source of this type of noise. Fortunately, although the resulting signal is invariably strong, it is sufficiently different from the more normal GPR data to be easily distinguished. Also, the position of most power lines is sufficiently well spaced for the underlying GPR data not to be obscured by large numbers of signals. Fig. 1.5 has a typical overhead power line signal at c. 585 m distance. This appears to be at ground surface level but was actually registered as the radar passed under the overhead power line. Observation during survey will usually identify any signals from power lines.

Although much of the regulation concerning GPR has at its heart the prevention of radars interfering with other equipment, it is more common to find that other transmitting devices interfere with the radars (see also

Chapter 15: Regulation). This is for two reasons, namely that GPRs are low-powered devices and their radiation pattern is therefore limited and secondly the screening generally prevents interference with equipment that is located anywhere other than directly below them. The same is not true of other transmitting equipment such as planes coming in to land, mobile phones used in the immediate vicinity, microphones, or person-to-person radio transmissions, all of which have the potential to interfere with the GPR data. Generally speaking, the frequencies of the other transmitters are different to those of the radar which means that they can be identified as coming from an external source. However, if there is any great density of signals, the result can be that the subsurface data is obscured by the interference from these external sources. The GPR operator is strongly advised to leave their mobile in a safe location away from the GPR while surveying.

There is another more subtle potential problem in secondary transmissions which can occur when the signals reflected by a target pass into the air and are re-reflected downward by an aboveground target. For this to cause an impact on the data, the secondary reflection tends to originate with strongly reflective material such as overhead metallic beams in a factory environment. The clue to this type of signal generally lies in the regularity of the patterning. Overhead metallic beams, for example, are usually spaced at regular intervals. A repeat of this patterning within the radar data would be a major clue that secondary reflections from overhead material were present.

SENSITIVE SITES

There are certain categories of survey site which are considered sensitive and for which the GPR surveyor is obliged to take care that his/her equipment does not interfere with other transmissions. This normally includes advising the relevant organization of the frequencies and the time at which the equipment is to be used. Although, as explained above, the low power used in GPR and the provision of shielding around the antennas reduce the risk of the radar causing interference, the transmissions at these sites are considered to be sufficiently critical for them to be included in the European guidelines, originally put together by the European GPR Association, subsequently updated for publication by the European Telecommunications Standards Institute (ETSI Guide EG 202 730, see also Chapter 15: Regulation).

The sensitive sites (sensitive radio environments) are:

- Airfields
- Prisons
- Defence Establishments
- Radio-Astronomy sites.

From practical experience, to this list should also be added railways since the railway operators usually have concerns about interference with the signaling systems, and hospitals, particularly intensive care wards. This does not mean that GPR cannot be used in these locations. It does mean that permission to operate must be obtained in advance and that the frequency bands that the survey team intend to use should be communicated in advance to those responsible for transmissions on the sensitive sites.

For all sites, not just the sensitive ones, the European Code of Practice requires GPR operators to ensure that undesired radio emissions are kept to a minimum through the use of properly regulated equipment which should only be switched on while survey measurements are being taken. GPR operators are also required to keep their equipment ground coupled to a layer of material of sufficient thickness and suitable properties for the signals to be fully absorbed and not scattered into the atmosphere. The equipment must have a deactivation mechanism so that it is not in constant use.

DETECTION OF AIR AND OTHER ABOVEGROUND SURFACE SIGNALS

Any radar signal generated from objects or reflections above ground will have traveled through air rather than through the soil. Where this is suspected, the data should be checked for curves or angular lines which can be fitted to a transmission velocity of 0.3 m/ns.

Note that, under certain circumstances, it is also possible to generate air signals from below the radar antenna. Fig. 9.6 shows two very different types

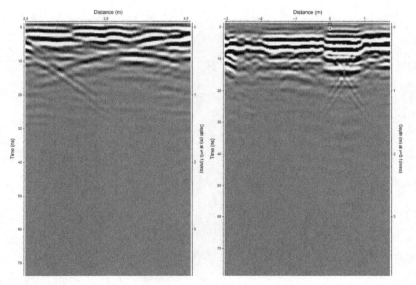

FIGURE 9.6   Examples of air signals.

of air signal, neither of which relates to aboveground objects. The first example, on the left-hand side, shows an air signal, at the beginning of the survey line, generated by the radar crossing a step within a building. As the radar reaches the edge of the step, there is an air gap between the base of the radar and the floor adjacent to the step. In the second example, there are two air reflections just after the marker on the data. These two signals emanate from inside an object depicted by two parallel layers. In this case the object is a grave, effectively a stone container for a coffin but for which the primary content is air. This type of secondary reflection is typical of vaults, crypts, cellars, and other subsurface constructions (cf Martinaud and Frappa, 2004).

REFERENCES

Martinaud, M., Frappa, M., 2004. GPR signals for the understanding of the shape and filling of man-made underground masonry. In: Slob, E., Yarovoy, A., Rhebergen, J. (Eds.), Proceedings of the Tenth International Conference on Ground Penetrating Radar. Delft, The Netherlands, vol. 2, pp. 439–442. ISBN 90-9017959-3.

ETSI Guide EG 202 730 Electromagnetic compatibility and radio spectrum matters (ERM); Code of Practice in respect of the control, use and application of ground probing radar (GPR) and wall probing radar (WPR) systems and equipment, ETSI, Sophia Antipolos, France. Available either from: <www.etsi.org> or <www.eurogpr.org>. Introduction to GPR—Code of Practice.

Chapter 10

Three-Dimensional Data

Chapter Outline

For each pass of the radar across the survey area, an image is generated which shows the depth of each target detected plotted in its position relative to the distance along the path traveled by the radar. As we have seen in previous chapters, although this is a vertical view equivalent to a vertical section, it has to be interpreted since it does not resemble a vertical section that might be drawn by a person, for example. GPR is not an optical method of detection and the images it produces are not comparable to those of optical devices. For many GPR applications, it is useful to be able to examine the data across the horizontal plane as well as in the vertical plane. For this purpose, the full data set can be combined into a three-dimensional grouping from which the patterning can then be examined from different angles, including directly above, as though looking down on to a horizontal surface. The location information gathered during the survey from a grid, GPS, or total station is used to combine individual radargrams into a three-dimensional block.

FORMATION OF THREE-DIMENSIONAL DATA

The usual way of doing this depends to a limited extent on which radar is being used as well as the amount of data gathered. Most makes of GPR have automatic software which will form three-dimensional data sets, at least for small sites or specific types of sites. As the position of the data is predetermined, the operating software can translate the data set into a three-dimensional cube, which can then be viewed in the horizontal plane as well as the vertical. This chapter is not concerned with the automatic processing of data into a three-dimensional data set nor with the capabilities of any particular software package. What follows is general guidance on the completion of a three-dimensional data set from a set of two-dimensional data lines.

Ground Penetrating Radar. DOI: http://dx.doi.org/10.1016/B978-0-08-102216-0.00010-2

It is not uncommon for larger sites to require postprocessing to form the three-dimensional data block. Either a survey grid or GPS can be used to determine the position of the individual lines. If the survey data were gathered using a grid and the radar started recording on the same initial reference line, then the data set will be already aligned along that starting line. If recording the data was started before the initial survey reference line, the markers added to the data which indicate the position of the starting line are used to align the data. If the data were collected by meandering, i.e., the first line in one direction and the second in the opposite direction, then the X-flip function will be needed to align the data so that all survey lines lie in the same direction.

Care must be taken when aligning the data set to ensure that the direction chosen to display the data from left to right across the screen matches the desired horizontal format to be extracted from the three-dimensional data. It is possible to arrange the data so that the view is inverted, resulting in incorrect location and possibly prejudicing interpretation of the image. This also depends on how the x- and y axes are arranged for data display in the software package.

For example, if the data have all been collected meandering along a North/South axis, the first decision is which way should we display the data: North to South or South to North? If we select North to South, then East will be at the top of the screen and West at the foot. Alternatively, if we select South to North, then West will be at the top and East at the bottom. It is important to know the orientation of the axes in the three-dimensional data set so that this matches the physical position on the ground (see Fig. 10.1).

For both parts of Fig. 10.1, the x-direction is the line taken by the radar and the y-axis indicates the relative positions of the various survey lines.

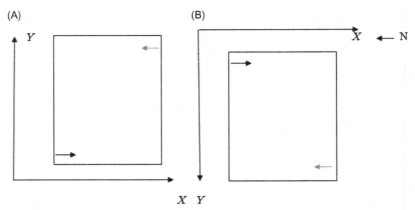

FIGURE 10.1 Planning the orientation of the two-dimensional data to fit the desired pattern of time slices extracted from the three-dimensional data block.

In Fig. 10.1A, if we assume that the black arrow represents line 1 of the survey, then the y values for each line will follow the same order as the survey lines themselves. So, if we assume that the transect spacing was 0.5 m and that line 1 was positioned at 1 m, the y values will follow the sequence: Line 1, $y = 1$ m; Line 2, $y = 1.5$ m; Line 3, $y = 2$ m; and so on. Provided that the software package arranges its x- and y axes in the same manner as Fig. 10.1A, this will be correctly displayed. However, if the axes are displayed as in Fig. 10.1B, the lines in the three-dimensional data block will be incorrectly placed and any horizontal image will have been inverted. Not only will this result in incorrect location information, there is also a risk that it may make a difference to the way in which the data patterning is interpreted although this depends to some extent on the type of GPR application.

Now consider a different start point and direction for the data collection. Assume that our first line of data follows the start point and direction of the gray arrow in Fig. 10.1A. It is obvious that, in order to get the location of any subsurface features correct, the direction of this line will need to be reversed by using an X-flip type function on the two-dimensional data in order to display the data from North to South as required. However, it should also be obvious that the y values will again depend on the format of the x- and y axes. For Fig. 10.1A, line 1 will have a higher y-value than any of the other survey lines. The $y = 1$ m will, in this case, apply to the last of the survey lines, not the first. If this correction is not done then the survey lines will be placed at the opposite end of the survey area from where they were actually recorded.

The other point to notice in this placement of survey lines is that, for our example, all of them lie parallel to the x-axis. This means that their start and end points have the same y-value. For orthogonal data sets, the process of placement is exactly the same except that the direction of the radar will lie parallel to the y-axis rather than the x-axis. For this case, it is the x placement that requires to be determined in order to place all survey lines in the correct location relative to one another. As all of these survey lines are parallel to the y-axis, the start and end of the lines will have the same value of x.

LINE PLACEMENT EXERCISE

Using Fig. 10.1B and assuming that the y-value at the position of the black arrow is 1 m, the transect spacing is 0.5 m and there were a total of 50 parallel survey lines completed, what should the y-value of the gray line be? What other processes, if any, have to be applied to the data? You may assume that data collection was carried out along alternate directions, beginning in the same direction and location as the black arrow, finishing along the line of the gray arrow. Readers should attempt this exercise before finishing the rest of the chapter.

It is usually possible to include both sets of orthogonal data into the one data file. The first set entered as running along the x-direction with a constant value of y for each file. The second set is entered as running along the y-direction with a constant value of x for each file. However, this is not always an advantage as depending on the transect spacing between lines and also the way in which adjacent data points are handled by the processing package, it can distort the data. It is sometimes easier to work from two orthogonal data cubes rather than a single one. A judgement has to be taken depending on the data set.

TIME SLICES

Horizontal data views extracted from the three-dimensional data set are called time slices (Fig. 10.2). This is because they are extracted at a constant time. This can be translated into a depth in centimeters/meters or feet/inches provided that the transmission velocity has been calibrated. Since transmission velocity may vary across the site or even with depth, the depth in nanoseconds is an accurate measurement but the depth in centimeters/meters (or inches/feet) will only be accurate if the correct velocity has been applied. If either of those conditions does not exist, the accuracy of the depth in nanoseconds is unaffected but the translated depth will not be accurate.

Fig. 10.2 is a relatively simple example where it is likely that the transmission velocity may not vary significantly. Depending on the site conditions and the complications of the data set, it may be necessary to construct a velocity file which applies a different calibrated velocity to different areas of the site. The extreme example of this is where either air or water exists in the subsurface.

In order to appreciate the importance of this, take the example of an empty buried cellar below the ruins of a demolished building. If we assume that the site is relatively dry with a calibrated transmission velocity of 0.095 m/ns and translate all depths at this velocity, a depth of 40 ns will equate to 1.9 m across the site. In the area of the cellar this is clearly incorrect once we reach the level of the air content of the buried cellar. At this depth and in this location, the transmission velocity increases to 0.3 m/ns.

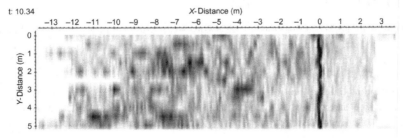

FIGURE 10.2 A simple example of a Time Slice showing the line of a buried pipe at c. 0.5 m (c. 10 ns) depth.

The equivalent depth of 40 ns can only be calculated if we measure the depth below ground surface of the top and bottom of the cellar. Assume for the purposes of illustration that the top of the cellar is 0.5 m below ground and the cellar measures 2 m from top to bottom. The first 0.5 m of ground is equivalent to 10.53 ns (to 2 decimal places). The next 2 m of ground will be crossed at 0.3 m/ns which is equivalent to 13.33 ns (to 2 decimal places). So the floor of the cellar will be reached in (10.53 + 13.33)ns or 23.86 ns. This leaves a further 16.14 ns to reach the full probing depth of 40 ns. Below the cellar, we can reasonably assume that the radio waves have returned to the calibrated ground transmission velocity of 0.095 m/ns. This means that, by 40 ns, the signals will have traversed a further 0.77 m. So the time slice extracted at 40 ns shows the majority of the subsurface at a depth of 1.9 m but the area containing the cellar, because of the effect of the air void, is actually reaching down to a depth of 3.27 m. The 40 ns is a constant but the depth in meters or feet is not, because of the variable velocity of transmission. In this example, the time slice could be said to be a quasi-horizontal view. Similarly, if there is water present in part of the site but not the whole site, the depth of time slices will be a constant in nanoseconds time but will be less in terms of meters/feet in the wetter areas due to the slower velocity of transmission.

One method of tackling this is to build a velocity file to translate the different areas of the site appropriately. This may or may not be possible depending on the extent to which transmission velocity varies. The example given is relatively straightforward as it is possible for transmission velocity to vary across a site as well as with depth.

For the example given above, the reader is advised to carry out the calculations independently but these do now follow.

The calculations for the example of the buried empty cellar are as follows:

1. 40 ns is the two-way travel time for the signal to reach the target and return to the receiver. Use speed (v) = distance$(2d)$/time (t). Distance is stated as $2d$ since the depth is equivalent to half of the distance below the surface which has to be traveled for the signal to return to the receiver. Depth, $d = \frac{1}{2}(vt) = \frac{1}{2}(0.095 \times 40)$ or 1.9 m.
2. Using the same formula, the first 0.5 m of depth is reached in a time of $(2 \times 0.5)/0.095$ or 10.53 ns to 2 decimal places.
3. The air gap in the cellar is 2 m (or 4 m of two-way travel time). Again, $v = 2d/t$ or $t = 2d/v = 4/0.3$ because of the increased transmission speed through the air. This gives a time of 13.33 ns to 2 decimal places.
4. The total amount of time used to penetrate to the floor of the cellar is 10.53 ns + 13.33 ns or 23.86 ns. This leaves $(40-23.86)$ ns or 16.14 ns of signal penetration left to reach the full 40 ns depth.
5. Exactly the same formula is applied again, this time in the form $2d = vt = 0.095 \times 16.14 = 1.533$ ns. The extra depth, d, is therefore 0.77 m.

6. The full depth that 40 ns penetrates in the area of the cellar = 0.5 m to the top of the cellar + the 2 m depth of the cellar + the extra 0.77 m, a total of 3.27 m.

The importance of time slices is the patterning which they reveal. Examples of where this is useful include the line of a pipe or cable, the interconnections of different pipes, the layout of an archeological site, the location of graves (clandestine or legitimate), the position of subsurface obstacles which need to be removed prior to development of the site, the position of reinforcement in concrete, the outline of a weapon or listening device concealed in a wall. The advantage of the time slices is that they effectively compare each survey line with all of the others so that the differences which appear in the horizontal (or quasi-horizontal) view signal clearly an outline of anomalous material. Those outlines can be helpful in determining the nature of the anomalous material. Unlike other geophysical methods, the features visible in the GPR data can be viewed by comparative depth which can also be helpful in determining their nature, if not always their relative dates of deposition. The only potential drawback is the processing time. However, it is worth noting that, with experience, this can be relatively rapid.

One additional feature worth noting is that time slices will reveal the angle of objects relative to the travel path of the radar. Although this may not matter, it can be useful when applying migration to the data if the linear feature being crossed by the radar does not lie at right angles to the radar's travel path. The angle can be measured from the time slice and applied in the migration to refine the result.

It should not be imagined that all three-dimensional materials are instantly recognizable, however, from its outline. In Fig. 10.3, the line of the power cable is easily identifiable but it is not clear from their outline what the former building remains represent.

The GPR is always correct in the variations that it picks up. The GPR data interpreter may or may not be as accurate in understanding the data. One of the ways of improving the quality of the interpretation is to examine the series of time slices and cross check major anomalies visible in the three-dimensional data with the corresponding two-dimensional radar plots. Although this can be time consuming, it is often a much more revealing exercise in terms of understanding the subsurface structures from which the anomalous signals are emanating.

Time slices are usually extracted from the three-dimensional data set on the basis of where changes occur in the patterning of the material distribution. Although some data interpreters prefer to extract slices of several nanoseconds thickness and others prefer to extract time slices at predetermined intervals, there is no hard and fast rule. The important points to observe are that the data illustrate a significant pattern and that this pattern has been interpreted in a consistent manner. Extracting at a single common depth in

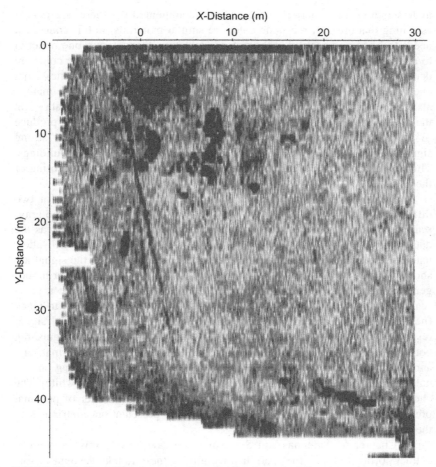

FIGURE 10.3 Time Slice from an urban site showing a buried power cable (including a branch) and former building remains.

nanoseconds may also be acceptable. Where the data have been collected over uneven ground it may be worth smoothing the data over one wavelength to compensate for the surface irregularities and the variation in subsurface depth which will result from these. Although it is usual to process the data in two-dimensional format before forming a three-dimensional data block, it is also possible to process the whole of a three-dimensional data block, should this be required.

In Chapter 3, More Fundamentals of GPR Operation, we looked at the impact of color and grayscale on the presentation of two-dimensional data. Many of the same considerations apply to time slices. Although color undoubtedly makes the data look more attractive and can assist the client in understanding the patterning observed and commented on, there is a risk of assuming that areas of the same color are similar materials. In this context, it needs to be remembered that the radar is mapping change from one material to another, not intrinsic material responses. Anomalies of the same color are demonstrating a similar level of change in electromagnetic properties from one material to the next. This does not mean that they are the same materials. Unless there is continuity of the signals either in the vertical data or in the time slice, there is no reason to suppose that the signals are the produce of an identical material change cf. Fig. 10.3 above where similarly strong signal returns indicate both utilities and the location of former buildings. This is something that would have to be tested, by trial excavation, coring or the equivalent.

If using grayscale for presentation of the data, there are options for two different types of presentation. Some GPR surveyors prefer to leave the grayscale as a spectrum from white through gray to black. In this format, as the data interpreter scrolls through the data, any anomalous material will alternate between black and white as positive and negative parts of the signal are shown. Where there is continuity in the electromagnetic properties (or homogeneous material), this will show as gray. The alternative is to use a grayscale that goes from black (positive signal) to white (zero crossing) to black (negative signal) and to envelope the signal so that all strong signal responses are indicated as being of high amplitude without distinguishing between the positive and negative aspects of the signal. In this format, any anomalous material will show as either dark gray or black, depending on signal strength and areas of relative homogeneity will show as white. The choice of grayscale and/or the use of color are usually a matter of personal preference. It makes sense to opt for whichever presentation contributes to the clarity of the results.

In Chapter 9, Antennas and Screening, we saw how it was possible to extract portions of data files and then recombine these to take account of surface obstacles. Fig. 10.4 shows a time slice from an urban site under redevelopment where parts of the surface were obscured by the addition of fencing to safeguard the equipment on site when not in use. The white area indicates the part of the site that could not be surveyed due to the position of the fence and the offset of the radar trolley from the fence. The top left-hand corner indicates the entrance to the equipment compound. There is no offset at this position because it was possible to move the gate. The purpose of the survey was to identify the position of a suspected former asphalt car park which can be clearly seen within the area of the fenced compound. Other strong signals, randomly distributed, relate to building material incorporated within the recent destruction of modern buildings on the site. The car park should have been destroyed as part of the site preparation but clearly had not been. The time slice allowed the feature to be located and removed prior to building work beginning on site.

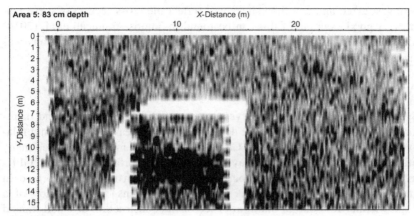

FIGURE 10.4   Time Slice from a Development site with surface obstacles.

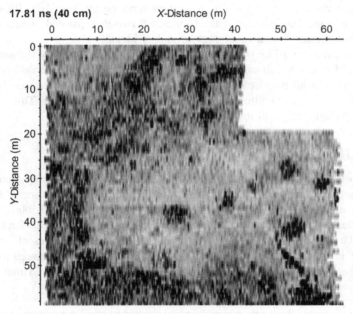

FIGURE 10.5   Archeological Time Slice illustrating the layout of a Bronze Age Cairn cemetery buried beneath upland peat in NW England (Utsi, 2007).

Time slices are particularly useful in archeological investigations. Fig. 10.5 demonstrates the mapping of a Bronze Age cairn site, buried beneath upland peat in NW England (Utsi, 2007). Note that it is not always necessary for time slices to be based on a rectangular grid pattern. The investigation of standing buildings, for example, may require columns or pillars to be investigated and any time slices will reflect the cross-section of these features. A good example of this can be found in Santos-Assunçao et al. (2014) where a comparative study of historical and modern building columns was made in order to understand their internal structure and condition.

CONTOURING

Typically GPR data treat the survey surface as though it was flat and therefore any depth measurements are depicted at depth measured below and relative to this notional flat surface. Generally this is not misleading since many sites investigated by GPR do have relatively flat surfaces. However GPRs are also used on slopes and mounds, features which may have formed in time after the buried subsurface. Where this happens, the formation of a slope whether it is due to build-up of ice or the construction of a soil mound (man-made or natural soil formation), it will not be easy to see which features are associated with each other on the basis of their depth unless the modern surface is corrected for its current topography. Fig. 10.6 illustrates contoured two-dimensional data. Contoured two-dimensional data can be made into a three-dimensional contoured data block.

Where a mound-like structure forms the modern surface, any flat surfaces concealed beneath will appear to dip since the depth below ground surface will increase with the rise of the mound. All GPR processing packages allow the data interpreter to adjust the data for topographic correction. This correction will then realign all the historic material in the subsurface.

AMPLITUDE MAPPING

It is possible to produce three-dimensional data which are made up from certain bands of amplitude so that the low responses are automatically eliminated from the image and only the signals of interest remain visible. This is particularly helpful for certain imaging purposes, e.g., forensic investigations. The key requirement is setting the limits of the amplitudes of interest, bearing in mind that these will always reflect the differences between materials rather than the intrinsic properties of any single material so that there

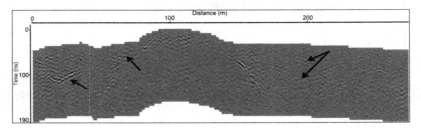

FIGURE 10.6   Example of contoured two-dimensional data (Bunting et al., 2014).

may be some variations even for similar materials. Fig. 11.1 in Chapter 11 is an example of this. The data interpreter has eliminated all but the strongest signal responses until a recognizable shape was formed by the items secreted inside the wall. The apparent distortion in the shape of the rectangular object is probably a result of a difference in packing materials at certain positions around the object so that the signal amplitudes in certain places are less in these areas.

ANSWER TO LINE PLACEMENT EXERCISE

There are 50 survey lines, placed at half meter intervals. From the orientation of the axes, line 1 lies along $y = 1$ m so lines 2, 3, and 4 will lie along $y = 1.5$ m, 2 m, and 2.5 m, respectively. Continuing the same sequence gives $y = 25.5$ m for line 50.

There is another process which requires to be carried out in order to place the lines correctly. Line 1 lies in the correct direction for consistency with the placement of the x- and y axes. Line 2, being on the opposite direction, does not. All of the even numbered survey lines from 2 to 50 will require to have their direction reversed through an X-flip function.

REFERENCES

Bunting, C., Branch, N., Robinson, S., Johnes, P., 2014. Ground Penetrating Radar as a Tool to improve heritage management of wetlands. In: Lambot, S., Giannopoulos, A., Pajewski, L., André, F., Slob, E., Craeye, C. (Eds.), Proceedings of the 15th International Conference on Ground Penetrating Radar, Brussels, Belgium.

Santos-Assunçao, S., Pérez-Gracia, G., Caselles, O., Clapés, J., Salinas, V., 2014. Geophysical exploration of columns in historical heritage buildings. In: Lambot, S., Giannopoulos, A., Pajewski, L., André, F., Slob, E., Craeye, C. (Eds.), Proceedings of the 15th International Conference on Ground Penetrating Radar, GPR 2014, Brussels, Belgium.

Utsi, E., 2007. Wetlands viewed through the antennas of a Ground Penetrating Radar. In: Barber, J., et al. (Ed.), Scottish Wetland Archaeology Project, Archaeology from the wetlands: recent perspectives, Proceedings of the 11th WARP Conference, Society of Antiquaries, Edinburgh, 2005, pp. 213−220.

Chapter 11

Common GPR Applications

Chapter Outline

GPR can essentially be used to investigate anything buried beneath another surface. The main exception is anything concealed below metal since radio waves cannot penetrate metal. Provided that the target is

- different in electromagnetic properties from its immediate environment,
- its depth and size is matched by an appropriate frequency of antenna,
- the target is not obscured by other similar material within a wavelength's distance,
- the soil is suited to GPR survey, and
- excessive moisture has not changed the electromagnetic response of the soil to the point where it is no longer suited to GPR survey,

then GPR is an efficient and reliable method of detection and there is little likelihood of the target(s) remaining invisible to the radar.

As a result, GPR is used for a very wide range of applications. Although, as might be expected, many applications are associated with specific frequencies of GPR, it does not do to be too specific, if only because exceptions can and do occur. In practical terms, it is extremely important to think in terms of what wavelengths are the best suited for the investigation being planned rather than blindly choose either whichever radar is readily available or what was used on previous applications of a similar nature.

SECURITY

Security sweeps tend to be searches for small targets often concealed close to the surface and are, therefore, most suited to high or very high frequencies of antenna (3–6 GHz). GPRs are used to look for unwanted objects concealed

Ground Penetrating Radar. DOI: http://dx.doi.org/10.1016/B978-0-08-102216-0.00011-4

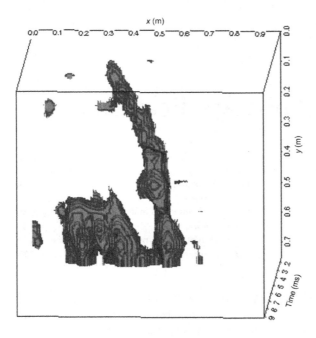

FIGURE 11.1 Three-dimensional image (4 GHz data) of a concealed weapon and another rectangular object (Utsi, 2010).

within walls, objects placed on the other side of walls and also for suitable spaces within walls, which are clear of potential obstructions such as power cables. Fig. 11.1 shows a three-dimensional image of a concealed weapon within a wall. This image was built up by carrying out a series of parallel survey sweeps of the wall then building up a three-dimensional data block before blocking out all signal returns below a selected high amplitude threshold. The threshold ensures that only the material which is very different from the normal contents of the wall is visible. Note that the metallic parts of the gun show up as a darker color than the remainder. These darker areas indicate a greater difference in the material in electromagnetic terms. The adjacent rectangular object (to the right of the weapon) appears also to have metal content.

Where security relates to the concealment of valuable or illegally held materials (see also Section "Mining"), this often concerns outdoor locations so that lower frequencies in the 1 GHz to 400 MHz range may also be used. As with all outdoor work, the choice of frequency depends on how deep it is necessary to probe, the size of the target(s), and the electromagnetic nature of the site environment.

STRUCTURAL ENGINEERING

Man-made structures whether modern or of historical significance usually require both quality control of the build standard and then monitoring for the development of problems throughout their economic life. Structural engineering surveys tend to use high-frequency antennas since the information sought is usually within 1 m (or 3 feet) of the accessible surface. Typical frequencies of GPR antenna used for this purpose range from the very high (4 and 6 GHz) down to around 1 GHz depending on the depth range required and any other limitations such as the spacing of the rebars in reinforced concrete (see also Chapter 12: Examples of Practical Problems). It is very rare to use lower frequencies, particularly where reinforced concrete is used, as a short wavelength is required in order for the electromagnetic pulses to pass through the space between the rebars.

Typical investigations include the detection of dowels, reinforcement bars and beams, tendon ducts, utility cables, the development of delamination behind floors, ceilings, internal and external walls, mapping of voids and cracking, and verification of tunnel wall quality. Allied to these structural surveys, it is not unusual to carry out examinations of land around and within the area of the structures, possibly using low-frequency antennas, particularly for industrial buildings or those with historical significance facing an environmental threat such as the impact from earthquakes. Any standing structure is suitable for GPR survey provided that there is no metal barrier between the radar and its targets and that the structure is not heavily impregnated with lossy materials, for example, a build-up of salt on coastal sites.

A closely related area of investigation is that of reinforced concrete whether part of a structure or part of its immediate environment or access road. The object is usually to determine the position and quality of reinforcement in the concrete but may also be linked to void detection, utility detection, or the identification of subsurface features such as underground waterways, depending on the nature of the site.

ROADS AND BRIDGE DECK

GPRs are used systematically on roads and bridge decks in many countries in order to check that their composition complies with the original design, to monitor their internal integrity for maintenance purposes or to identify the reasons for underlying problems impacting the surface. Although different frequencies of antenna may be used for these varied purposes, there is some overlap in objectives and they often form part of the same investigation. As with structural investigations, utility detection and verification of reinforced concrete may also form part of the same project. The surveys may also carried out because the original build documentation is missing or incomplete. Another speciality of road investigations lies in the investigation of surface cracking, for which see Section "Crack Detection".

Determining layer thickness is a principal objective of this type of investigation which means that antenna frequency is critical. In order to determine layer thickness accurately the antenna must be capable of resolving the top and bottom of all subsurface layers. Typically therefore more than one frequency of antenna is likely to be used. Commonly used antennas range from the medium frequencies of 400–500 MHz, used to provide an overall view to the full depth required to high-frequency antennas in the range of 1–2 GHz, used to provide greater detail. Since the information is normally generated simultaneously, it is usual practice to use as multichannel GPR for this purpose. Fig. 11.2 shows an example of 400 MHz and 1.5 GHz data taken along a stretch of road, which contains an area where repeated repairs have been carried out.

The top image is taken from the 400 MHz data and the lower image from 1.5 GHz data (1 GHz being 1000 MHz). Note the difference in resolution of the layering in the repaired section. The lower image shows much greater detail of this section and the upper image the repair within the context of the full probing depth. It is not difficult (and should not be difficult) to reconcile the two data sets.

Delamination of the wearing course involves the detection of small air voids close to the surface and therefore requires very high-frequency antennas (greater than 3 GHz) to be effective (Carrick Utsi et al., 2014).

The speed with which road and bridge deck investigations are carried out varies from one country to another. For example, the preference in the United Kingdom is for high-speed investigation in order to keep traffic congestion to a minimum on otherwise crowded roads. This requires a fast acting GPR, mounted on a vehicle. In the United States, however, speed is not always a requirement and although whole or partial road closures are not welcomed, they are more acceptable to the Highways Authorities. In all countries, some road closure or traffic management is to be expected since it will sometimes be necessary to complete orthogonal survey lines across the

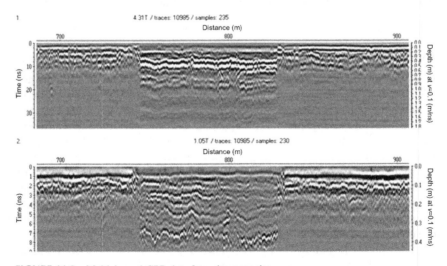

FIGURE 11.2 Multichannel GPR data from the same site.

road as well as longitudinal surveys. The orthogonal survey lines are usually collected by a GPR surveyor on foot, using a trolley mounted system but are equally likely to use more than one frequency of antenna.

The GPR data are collected in lines along the road at regular intervals which are likely to correspond to predetermined spacings such as the near-side wheel track, center lane, and offside wheel track as a minimum. The equipment may be suspended below the vehicle or attached at the rear. An odometer (or encoder wheel) can be used for distance measurement with or without GPS. In some instances, one of the vehicle's own wheels may be used as the odometer for the GPR system as well as the vehicle. Orthogonal surveys are more likely to follow a grid pattern, the spacings between adjacent survey lines being dictated by the target definition required.

A commonly recurring problem with road surveys is the presence of standing water on the road surface, which is likely to adversely affect the results. In the winter season, the use of salt as a defrosting agent may result in salt water coverage of the road, in which case signal attenuation is a sufficiently major impediment for a GPR survey to be inadvisable. Detailed guidance on GPR Pavement survey is available from the European GPR Association website.

AIRPORT MAINTENANCE

There are a number of aspects to airport maintenance, which include both monitoring the considerable daily wear and tear on runways and taxiways but also the critical element of detecting underground utility pipes and cables, which include everything from fuel lines to water, electricity, lighting, and data carriers, often in close proximity to one another. Given that most airports are surrounded by businesses and homes, the potential damage, which could result from an accident to one or more of these pipes and services, tends to be enormous, making accurate, and comprehensive utility detection critical.

The constraints in timing of airport maintenance surveys are slightly different depending on whether the airport is civil or military. For example, it is often possible to gain access to military airports during the day, depending on the scheduled activities. This is rarely the case for civil airports where most investigations have to take place overnight in designated areas in order not to disrupt the flight schedules. Speed of data collection is often an issue for similar reasons, i.e., to minimize the impact on the airport's main activities. Against this, it is also extremely important to recognize that accurate location is critical in view of the potential hazard to life should a breach occur of an undetected utility.

One potential pitfall to look out for is the use of appropriate frequencies of antenna. Runways in particular are usually heavily reinforced. It has happened that very closely spaced rebars have been mistaken for a stratigraphic layer change, e.g., the boundary between the upper levels of asphalt and the underlying concrete. Unless the antenna frequency is high enough for the radar to distinguish each rebar from its neighbors on either side, the closely spaced rebars will appear to form one continuous layer which, unless coring is carried out, is likely to be misunderstood as the interface between asphalt and concrete. As discussed in Chapter 2, Wavelengths and Why They Matter, this means that the wavelengths emitted by the GPR must be equal in length to or shorter than the spacing between the rebars. In this particular example, one result was that no pattern of reinforcement was detected and it, therefore, looked as though the concrete had not been reinforced. This is sufficiently unlikely to give the clue that the high-frequency antenna used was still too low for the survey requirement and that a high-frequency antenna was required. The high-frequency antenna then revealed the close positioning of the reinforcement bars.

It is evident from the foregoing that most if not all GPR surveys of airports will require a minimum of two antennas, one of which must be high frequency. If full lane coverage is required, multiples of these antennas can be used for this purpose. If speed is considered essential, then either a fast acting multichannel GPR, preferably mounted on a vehicle, is required unless several repeat survey sweeps are made instead. Understanding wavelengths is critical to carrying out a good airport investigation because of the limitations imposed by reinforced concrete. As for road and bridge investigations, it will also be necessary to carry out orthogonal surveys for which some form of closure or traffic control will be necessary.

CRACK DETECTION

Both roads and airport surfaces suffer from cracking. In the case of the roads, it is the depth of the cracking from the surface down which matters in deciding whether to resurface or to strip back and lay an entirely new surface. This is top-down cracking. Although it has been claimed that top-down cracking can be measured using traditional GPR antennas, this is largely a hit and miss affair, particularly if the survey is carried out at speed. There are a limited number of specially adapted (cross dipole) antennas available for this task. Cross dipole antennas favor linear features over layer information and can, therefore, be used for measuring individual crack depths (Forest and Utsi, 2004). If traditional parallel polarized antennas are used, although it may be possible to distinguish areas of cracking as opposed to individual crack depths, detection of all targets is unlikely to be successful. Antennas used require to be high or very high frequency, usually the latter since cracks are near surface small features.

The situation is slightly different for airport surfaces. Structural engineers assess the top-down cracking on the basis of visual inspection but have no nondestructive means other than GPR for assessing the bottom-up cracking. Coring is both intrusive and damaging with the potential for additional damage from the creation of foreign object debris on the surface, which is potentially dangerous to aircraft engines. This type of cracking typically forms at the position of the joints in the subasphalt concrete and permeates up into the asphalt above. As for the roads, the decision lies between stripping back and then replacing the asphalt or adding another top layer to the existing structure. This, in turn, depends upon the degree of top-down cracking to bottom-up cracking. As far as the GPR is concerned, provided that suitably adapted antennas are used, there is no difference in detecting top-down or bottom-up cracking since both form acceptable targets. Crack detection GPR can be used to map the position of the joints, to determine whether cracks have permeated into the asphalt layer and to map these detected cracks and any other anomalies, including localized delamination (Birtwisle and Utsi, 2008; Utsi et al., 2008).

The major drawback is that this is not an investigation that can be carried out at speed. It is, therefore, not possible to complete the layer survey at the same time as the crack detection and measurement investigation. The latter needs to be focused on the critical areas thought to be most at risk.

For the detection of areas of generalized cracking (as opposed to detection and measurement of individual cracks), often used in the protection of historical buildings, see the section on Archeological surveys.

UTILITY DETECTION

The largest single use of GPR is for utility detection, normally in advance of either construction or the burial of additional services below the ground surface. Although it is not the only means of detecting pipes and services, it is a crucial tool because electromagnetic location (EML, e.g., "cat and genny") cannot be used in all circumstances. Unlike the electromagnetic methods, GPR can detect buried plastic pipes and operates effectively within the vicinity of electrical substations. It is, therefore, a good complement to the EML. However the two methods differ in their depth measurements since GPR measures to the top of the buried utility and EML to the center.

The reason why utility detection forms such a large proportion of GPR survey work is the lack of complete information as to where pipes and cables are buried, both in position relative to above-ground features and depth. This has arisen from records being lost or destroyed, from above-ground developments which have significantly changed the reference points used to record the location, from schematic representations of the location and, worst of all from mis-transcription from one system of measurement to another (e.g., feet and inches to meters). It is a worldwide problem. The other part of the equation is the risk to human life and well-being caused if a cable strike or pipe burst occurs, particularly if electric cables are involved. It is, therefore, mandatory in most countries to use some form of utility location in advance of ground disturbance.

GPR is not always used as efficiently as it could be and the emphasis is often on limiting the cost of the survey for the short-term benefit of the client. This frequently results in limitation of the amount of coverage rather than considering the long-term costs and consequences of nondetection, which is, after all, extremely dangerous and usually potentially life threatening. This means that utility surveys vary considerably in their scope even down to whether GPR is used at all. In recent years, a number of organizations have laid out specifications for utility surveys with different levels of confidence so that there is at least some understanding of what is being commissioned and provided. The American Society of Civil Engineers were the first to take this route with the introduction in 2002 of their "Standard Guideline for the Collection and Depiction of Existing Subsurface Utility Data," which has also gone on to provide the foundation for many of the other equivalent documents. In the United Kingdom, The Survey Association (TSA) produced its "Essential Guide to Utility Surveys" in November, 2009, guidance which has subsequently been updated to a fourth version. This has been mostly superseded by the British Standards Institution's introduction of its Publicly Available Specification on utility detection, known as PAS 128 (2014). This also provides a recognizable connection between the cost of the survey and the deliverables gained from it. Guidance on the use of GPR for utility detection is also available from the European GPR Association.

There are three attributes of GPR, which are perceived to cause problems. First, detection depends upon crossing the pipe or service, not on following it. This implies a need for more time spent on detection than with EML, particularly in an area where pipes have been laid in different orientations. A full image of where the various pipes lie and which one joins to which other would require survey in two orthogonal directions, effectively doubling the survey time. Secondly, GPR has a limited ability to distinguish adjacent utilities in an environment where it is increasingly common to lay pipes and cables as close to one another as possible in order to cope with the many demands of modern services, including electricity, water, telephone, broadband capability, and the disposal of foul water (cf Fig. 2.2, Chapter 2: Wavelengths and Why They Matter). Utilities, which lie close to one another relative to the wavelength emitted by the radar, are not individually distinguishable. This particular attribute is not as major a problem as it can be tackled by using high frequency as well as low-frequency antennas and by judicious use of electromagnetic locators alongside GPR. Another complementary technique sometimes used is vacuum extraction, which exposes the buried utilities after detection, making ground truthing possible.

Finally, there is the question of imaging potential. Utility surveyors are often reluctant to spend time completing an area survey, preferring instead to record a series of relatively widely spaced survey lines. Although this undoubtedly speeds up the process of data collection and potentially reduces the cost, it should be remembered that the radar does not and cannot perform as an optical device. Even if it could, the spacing between adjacent survey transects may be large enough to hide more than one pipe or interconnection. Any requirement for full data collection implies a need for closely spaced survey lines.

There is no reason why GPRs cannot be used strategically, i.e., close survey where the risk of nondetection by other methods is high, more widely spaced where the risk is judged to be low. Effectively, this is a judgment call. It is, therefore, extremely important that anyone carrying out utility detection is capable of considering the probability of needing full GPR data collection rather than blindly following the same practice applied on a previous site. The risks to life and limb in the case of a breached utility are too high for the GPR requirements to be ignored.

Although it is rare for GPR to be excluded from utility detection surveys, this has occasionally been proposed on the grounds that it is unnecessarily expensive and may not add value to the survey. The problem with this approach is that the surveyor (or client) is then assuming that every potential hazard buried below ground is detectable by EML. Plastic gas pipes have already been given as one example where detection by GPR is essential. Another example is where two or more high-voltage cables are deeply buried with the 50 Hz phase balanced to cancel out peaks and troughs in order to avoid unwanted electromagnetic effects such as heating or melting of the oil-filled cable insulators. Unless there is direct access (e.g., at a substation), EML will not work and detection by GPR becomes essential. It is extremely dangerous not to deploy GPR at all.

One other contentious issue is whether it is acceptable to physically mark up on site the position of all utilities, which can be observed on screen during the survey instead of postprocessing the data to test for any which remained unseen on screen. Although there are utility surveyors who consider that postprocessing represents a potential waste of their client's money, it is the general view of highly experienced GPR surveyors that this is potentially dangerous. There are many reasons why a particular utility may not be visible in real time in the data, including the possibility of being overshadowed by a stronger signal response from an adjacent pipe or other buried structure (cf Chapter 3: More Fundamentals of GPR Operation). The environment in which the utilities are placed can also reduce the visibility of pipes and cables, and the increased use of recycled materials can also have an obscuring effect. Add to this, the on-site problems of working with screens in either strong sunshine or driving rain and it is evident that there is at least a potential risk of not following up a survey with data processing to ensure as a high-detection rate as possible. Fig. 11.3 illustrates the point with data from a suburban utility detection survey. Only one pipe is clearly visible below the asphalt layer between 12 and 13 m distance but these data contain all the main domestic utilities. Not all pipes and cables are equally and immediately visible.

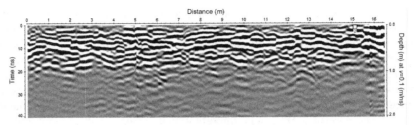

FIGURE 11.3 Not all utilities are immediately visible for marking up on site.

Although many utilities are placed relatively close to the ground surface, there is considerable variation, depending on the nature of the amenity being provided, in the depth at which they are traditionally laid. Although the utility industries have expressed concern about locating plastic pipes, accident statistics suggests that cable strikes are still common. This has frequency implications. Utility detection GPRs tend to be centered around 800 MHz, a compromise between using a 400/500 MHz unit and a 1 GHz. The smaller size of cabling and its proximity to the surface suggest that it might be better to deploy dual frequencies of 400/500 MHz and 1/1.5 GHz for improved detection. As noted above, this potentially also improves the surveyor's ability to differentiate one service from another.

There are a wide range of GPRs intended specifically for utility detection, both single and multichannel. At least one manufacturer produces antennas in two polarizations intended to speed up detection in two orthogonal directions. Although many pipes and cables lie under road surfaces, there are also quite a few along the margins of roads. Not all utility detection GPRs have the capability of tackling relatively uneven surfaces in this type of location.

ARCHEOLOGY

Although traditionally archeological geophysicists have predominantly made use of other geophysical methods, notably resistivity and magnetometry, GPR is increasingly being used for the detection and nondestructive investigation of archeological sites. It is also particularly useful in the examination of historical buildings. The main characteristic of archeological sites is their wide variety, covering everything from small-scale structural survey of buildings through urban redevelopment sites of varying sizes, wetland surveys on organic soils, investigation of abandoned rural sites, and even large-scale surveys across tracts of open countryside. Good examples include but are not limited to:

- the recent extensive exploration of the area around Stonehenge, England by Professor Vincent Gaffney's team (Gaffney et al., 2012),

- mapping of the former Abbey at Scone, Perthshire, Scotland currently being used to generate a three-dimensional virtual model of the building (O'Grady, 2008),
- the work at Audley End, Cambridgeshire, England by English Heritage (Linford, 2014), and
- The exploration of a 13th century mosaic and its associated burials in Westminster Abbey (Utsi, 2006).

The most commonly used antenna frequencies for archeological investigations are 250 and 400/500 MHz. Although these are frequently the most appropriate frequencies, especially on outdoor sites, there is a case to be made for the use of higher frequency investigations both indoors and outside. Currently, there is a perception that higher frequencies cannot be deployed outdoors although this is clearly not the case. Most archeological sites lie within the 1−2 m of the ground surface. Depending on the depth of groundwater and the amount of moisture contained within the ground, high-frequency antennas could well be useful, as they can be in forensic investigations (see below). Recent use of swept frequency radars to map sites in detail also supports this contention.

With respect to indoor investigations, a balance has to be struck between the frequency of antenna used, its depth capability and the target definition it can provide. It is not always appropriate to use a low-frequency antenna inside a small building, particularly if the extant remains are shallow and not particularly large, at least in their vertical extent. This could limit the ability of the antenna to detect the archeological remains. It is important to consider both wavelength and frequency in advance of the survey.

There are internationally accepted guidelines for the data collection, interpretation, and reporting of archeological geophysics, including the use of GPR, published by Historic England (formerly English Heritage) and also by the European Archaeological Consilium (English Heritage, 2008; Schmidt et al., 2015). These give guidance in detection by type of archeological feature as well as recommendations for optimal geophysical detection method. Useful as this undoubtedly is, as far as GPR is concerned, the reliance on categorizing by type of feature rather than focusing on the potential detectability of the feature relative to its environment (which can often vary from one site to another) is not always as useful as it might be. It is easier to assess the potential for detection by GPR in terms of the electromagnetic contrast the feature may make with its immediate environment. As an example, the European guidelines suggest that wetlands are a difficult environment for GPR. A more realistic assessment is to take into account (1) the slowing down of transmission velocity implied by water content and (2) the necessity for freshwater without the build-up of free ion content and, as discussed earlier, to consider the use of a low-frequency GPR.

The guidelines also make recommendations in respect of sampling strategies. All archeological data are partial by very nature whether this is an abandoned site, a redeveloped one or even a continuously occupied building (cf Rodwell, 2012). Increasing the data density has a marked effect on both the accuracy of the location of the remains and also the ease with which their patterning may be understood (Dennis, 2004).

There has been considerable interest in Mediterranean countries, particularly those whose historically important buildings are intermittently subject to earthquake activity, in the detection of cracks in historical buildings and GPR is being increasingly used for this purpose. This is not a question of measuring individual cracks but in the detection of areas of cracking and fault lines which weaken the building and put it at risk in the event of another earthquake.

Another interesting development is the use of swept frequency systems in the detection and analysis of archeological sites (cf Chapter 10: Three-Dimensional Data). Swept frequency systems sample through a range of frequencies instead of through time which, in theory at least, should effectively match the site to its optimal frequency of investigation (Linford et al., 2010).

It is not unusual for GPR data to be combined with the data from other geophysical methods. Recently, there has been interest in using good GPR data as the basis for re-creating buildings in virtual reality (O'Grady, 2008). This is done by using GPR data from an area survey for the ground plan of the building and combining this with any extant remains above ground, for example, fragments of stone from the remains of a window and any historical evidence available. It is obviously necessary to combine the expertise of a number of different specialists in order to achieve a virtual reconstruction. The exercise allows the viewer to enter a building which is no longer in existence, to view its size and shape, and even to gage the way in which light would have entered the building. This is a very exciting development.

FORENSICS

This is an unusual branch of GPR survey. Although it is not a frequent request, GPR can be extremely effective as a search method and is the only practical nondestructive method where a body has been buried beneath or within man-made building materials. Typical investigations are the detection of mass burial graves, searches for illegally buried missing persons, the detection of structures associated with clandestine burials, the discovery of concealed guns and other weapons, and also the detection of illegal dumps of potentially problematic material, for example, noxious chemicals (Groen et al., 2015). It may also be necessary to clear a site, i.e., to use GPR to search a site in order to establish that no illicit burials have been made (cf Utsi, 2013).

The range of equipment used is much the same as that deployed in archeological surveys and the work is frequently carried out alongside forensic

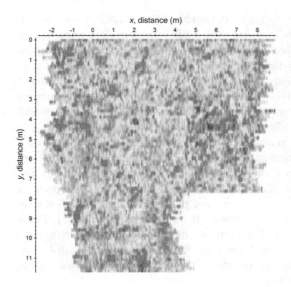

x, distance (m)

y, distance (m)

FIGURE 11.4 One of a series of Time Slices (horizontal view) showing potential hiding places for an illegally held firearm (1 GHz data).

archeologists. This type of investigation is only suitable for someone who is experienced in using GPR as strategic decisions will be necessary on the sampling strategy and there are invariably very strong legal and emotional reasons why the risk of failure to detect needs to be minimized. It is not a suitable GPR application for a beginner.

As with all GPR surveys, the size and depth of targets dictates which antenna frequencies are the most appropriate. The survey parameters and line spacing also matter as it is important to ensure that the target(s) will be reached by the radar signal at least once, preferable more. Fig. 11.4 shows a time slice (horizontal image) taken from a search for a buried (illegally held) firearm. Intelligence given to the police identified the area, which was open ground with associated woodland. In view of the relatively small target, a 1 GHz antenna was used to map potential hiding places. The strongest amplitude signals are delineated in bold in Fig. 11.4. The stronger the amplitude of the signal, the greater the difference between the ground and the anomaly. The strongest responses are likely to be from metal content, air gaps, or water. Since the site is dry, the strongest signals will indicate the presence of a metal object or air gaps. Many of the signals in Fig. 11.4 are voids (rabbit holes, very suitable for concealing a weapon), others are disturbed ground or tree roots. The smallest sized anomalies were ignored, as likely to be too small for the weapon. The larger areas of signal, particularly the linear features, were excavated by a search team in order to locate the weapon.

Clandestine burials are typically detectable by association, i.e., where there is a build-up of fluids or gases as a result of recent decomposition or where disturbance in the soil is still visible. It is the associated features, which act as targets rather than the bodies themselves. Bones are not normally detectable by GPR since they continue to interact with the soil around them so that, after a relatively short time, their electromagnetic properties are identical to their environment. For mass graves, there are usually residual outlines to the graves even after considerable time has passed.

There are no published statistics as to how long the signs of disturbance will remain in the soil although this is likely to depend on many factors including the depth of burial, the soil type, and the weather, especially the rainfall pattern after burial. There are a number of experimental projects monitoring the decomposition of human and pig remains from which it may be possible to extrapolate some guidelines (e.g., Schultz et al., 2002). There has been at least one recovery of the remains of a single missing person after a burial period of 14 years. The GPR data indicated a number of disturbed areas within the search site, one of which, on excavation, contained visible remains of the victim's clothing.

ENVIRONMENTAL AND GLACIOLOGY

This type of investigation normally uses low-frequency antennas because the information required is usually relatively deep. The environmental category also includes wetland site conservation and low-frequency antennas are necessary in this type of context in order to compensate for the slowing down effect of water (cf Chapter 4: The Effect of Water and Air). Environmental investigations are often to track particular strata whether this is the base of a peat layer or some other soil/rock. Tracking the movement of pollutants, in soils or in lake contexts, is another common environmental survey. The GPR data will usually be supplemented with cores taken in order to identify the material forming each of the strata. Coring used to be the only option for assessing the subsurface topography of sites of scientific interest. Using a GPR allows for a greater amount of continuous data to be collected in a shorter time. Coring can be a very time-consuming process and it provides information for where the core was extracted, nowhere else. The GPR data can be generated continuously across long distances and the results used to determine where cores are necessary for layer identification. Fig. 11.5 shows the profile of a classic peat basin from an investigation of a fully waterlogged peat site, one of the United Kingdom's largest remaining peat reserves, Bowness Common, Cumbria.

A similar logic applies in the case of glaciology studies. Although coring will still be necessary, it is possible to track stratigraphic information continuously over long distances using the core information to identify strata. Freshwater ice is particularly well suited to GPR. Not only do the electromagnetic pulses travel at a relatively rapid velocity (Table 5.1,

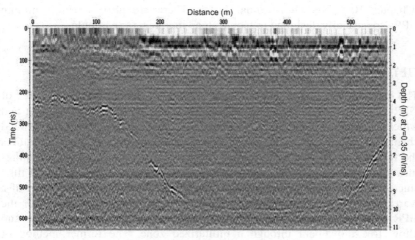

FIGURE 11.5   GPR definition of a classic peat basin (Utsi, 2001).

Chapter 5: Velocity Calibration) but this is a relatively low loss environment in which it is sometimes possible to obtain a greater depth than would otherwise be possible. It is not unusual for GPRs used in this type of environment to have depth settings, which are effectively in excess of 20 wavelengths. This also means that high-frequency GPRs can be used in this environment. A good example is the 2004 paper by Arcone and Laatsch where a 400 MHz antenna was used alongside the more traditional 100 MHz.

Glaciological targets range from the detection of potential subsurface hazards such as crevasses, investigating the structure and movement of glaciers to mapping of snow layers, and even location of bodies buried as a result of avalanche (e.g., Modroo and Olhoeft, 2004). The work is not always carried out on Earth, planetary exploration is another important branch of this type of investigation.

MINING

This category of GPR survey covers the detection of ores and mineral rich strata, the investigation of former mine workings which can present a hazard for modern mine working, particularly if water is present, and the assessment of spoil heaps. Delamination within specific strata can also present a risk to safety and GPR can be used to detect this (e.g., Triltsch, 2004). The depths involved generally mean that very low-frequency radars are the most suitable. However there are two other applications which may involve the use of higher frequencies. These are security sweeps for potential thefts of valuable material and in-mine measurements down to the mineral seam.

The terrain is usually challenging and varies from inside the mine itself to mountainous terrain or jungle conditions outside. Snake antennas are commonly used for this type of application and borehole antennas are also commonly used

(Chapter 10: Three-Dimensional Data). There are also specialist mining GPRs specifically designed to tackle steep inclines, for example.

DETECTION OF UNEXPLODED ORDNANCE

The detection of unexploded ordnance (UXO) is a highly specialized area of GPR expertise, which has been the subject of a great deal of research and which continues to evolve as the unexploded weapons left in the ground change in size, shape, and sophistication. Humanitarian demining requires the removal of all threats to human life and safety and is a higher requirement than military demining. Depending on the terrain, on the length of time since the mines were laid down and whether or not the area is still part of a war zone, UXO detection may be on foot, using hand-held devices or the GPR may be attached to a vehicle in order to protect the occupants and allow free movement through a militarized zone. The former method of investigation usually relies on a single channel GPR, the latter on a multi-channel. In both cases, it is possible to find systems which combine other geophysical techniques with GPR in order to improve detection rates.

The frequency or frequencies applied for this type of task vary with the formation and size of the targets planted. This makes historical unexploded ordnance easier to detect, at least in principle. Generally speaking, the depth at which unexploded ordnance is buried is relatively shallow in order to maximize its impact, which means that high frequencies or very high-frequency antennas are suitable but there are also circumstances where a very low-frequency antenna can be useful. This is also an application for which innovative antenna design is often used.

The principal problem is the potential for false positives, i.e., a positive reading for a nonlethal object and false negatives, i.e., missing a real target. Although it is tempting to think that a false positive represents no danger and therefore is a safe result, the reality is that a false positive or series of false positives leaves an unsafe impression in the mind of the GPR surveyor, which can potentially lead to a real target being either ignored or handled inappropriately with disastrous results. A lot of recent effort has, therefore, gone into being able to image all targets as accurately as possible.

There is an excellent summary of research into this area and some of the equipment and methods applied in Daniels (2004) and an excellent overview paper by the same author in Daniels et al. (2008).

REFERENCES

Arcone, S.A., Laatsch, J.E., 2004. Reversing the roles of high and low frequency to profile the dynamics of an ice shelf. In: Slob, E., Yarovoy, A., Rhebergen, J. (Eds) Proceedings of the 10th International Conference on Ground Penetrating Radar, Delft University of Technology, Delft, The Netherlands 2004, pp. 781–784.

ASCE38-02, 2002. Standard guideline for the collection and depiction of existing subsurface utility data. Available from www.asce.org.

Birtwisle, A., Utsi, E., 2008. The use of ground penetrating radar to detect vertical subsurface cracking in airport runways. In: Chignell, R.J., Rogers, C.J. (Eds) Proceedings of the 12th International Conference on Ground Penetrating Radar, University of Birmingham, Birmingham, UK.

Carrick Utsi, E., Birtwisle, A., Utsi, V., 2014. GPR analysis of bridge deck construction and the detection of delamination. In: Forde, M. (Ed.) Proceedings of the 15th International Structural Repair Conference, London, July 2014.

Daniels, D.J., 2004. Ground Penetrating Radar, second ed. The Institution of Electrical Engineers, London, pp. 501–624.

Daniels, D.J., Curtis, P., Lockwood, O., 2008. Classification of landmines using GPR. In: Proceedings of IEEE RadarCon2008, Rome.

Dennis, T.J., 2004. Imaging geophysical data – taking the viewer into account. Archaeol. Prospect. 11, 35–48.

English Heritage, 2008. Geophysical Survey in Archaeological Field Evaluation. English Heritage Publishing, Swindon.

EuroGPR developed Forensics Guidelines in consultation with the UK Forensic Science Regulator and their own experienced members. www.eurogpr.org/vn2/index.php/introduction-to-gpr/guidelines.

EuroGPR guidance on utility surveys can be found at www.eurogpr.org/vn2/index.php/introduction-to-gpr/guidelines. This also has a list of other relevant European documents.

EuroGPR offers guidelines on Pavement Structural Surveys at www.eurogpr.org/vn2/index.php/introduction-to-gpr/guidelines where additional reference material is also listed.

Forest, R., Utsi, V., 2004. Non-destructive crack depth measurements with ground penetrating radar. In: Slob, E., Yarovoy, A., Rhebergen, J. (Eds) Proceedings of the 10th International Conference on Ground Penetrating Radar, Delft University of Technology, Delft, The Netherlands, 2004, pp 799–802.

Gaffney, C., Gaffney, V., Neubauer, W., Baldwin, E., Chapman, H., Garwood, P., et al., 2012. The stonehenge hidden landscapes project. Archaeol. Prospect. 19 (2), 147–155.

Groen, W.J.M., Márquez-Grant, N., Janaway, R.C. (Eds.), 2015. Forensic Archaeology A Global Perspective. Wiley, Oxford.

Linford, N., 2014. Rapid processing of GPR time slices for data visualisation during field acquisition. In: Proceedings of 15th International GPR Conference, Universite Catholique de Louvain, Brussels, Belgium, pp. 731–735.

Linford, N., Linford, P., Martin, L., Payne, A., 2010. Stepped frequency ground- penetrating radar survey with a multi-element array antenna: results from field application on archaeological sites. Archaeol. Prospect. 17 (3), 187–198.

Modroo, J.J., Olhoeft, G.R., 2004. Avalanche rescue using ground penetrating radar. In: Slob, E., Yarovoy, A., Rhebergen, J. (Eds) Proceedings of the 10th International Conference on Ground Penetrating Radar, Delft University of Technology, Delft, The Netherlands, 2004, pp.785–788.

O'Grady, O.J.T., 2008. Tracing the medieval royal centre at scone, Medieval Archaeology, 52. Maney Publishers, Leeds, pp. 376–378.

PAS 128 (2014) Specification for underground utility detection, verification and location, British Standards Institution. Available from www.shop.bsigroup.com/.

Rodwell, W., 2012. The Archaeology of Churches. Amberley Publishing, Stroud, p. 249.

Schmidt, A., Linford, P., Linford, N., David, A., Gaffney, C., Sarris, A., et al., 2015. EAC Guidelines for the Use of Geophysics in Archaeology. EAC Guidelines 2. Europae Archaeologia Consilium, Namur, Belgium, ISBN 978-963-9911-73-4.

Schultz, J.J., Falsetti, A.B., Collins, M.E., Koppenjan, S.K., Warren, M.W., 2002. The detection of forensic burials in florida using GPR. In: Proceedings of the Ninth International Conference on Ground Penetrating Radar, University of California, Santa Barbara, CA, USA.

The Survey Association guidelines are available from http://www.tsa-uk.org.uk/for-clients/guidance-notes/.

Triltsch, G., 2004. Stepped frequency GPR field trials in potash mines. In: Proceedings of the 10th International Conference on Ground Penetrating Radar, Delft University of Technology, Delft, The Netherlands, 2004.

Utsi, E., 2001. The investigation of a peat moss using ground probing radar. In: Proceedings of the Workshop 'Remote Sensing by Low-Frequency Radars'. Consortium of Research on Advanced Remote Sensing Systems and the European Association of Remote Sensing Laboratories, Naples, Italy, September 2001.

Utsi, E., 2006. Improving definition: GPR investigations at Westminster Abbey. In: Daniels, J., Chen, C.C. (Eds) Proceedings of the 11th International Conference on Ground Penetrating Radar, Columbus, OH.

Utsi, E., 2010. GPR as an imaging device: some problems and potential solutions. In: Persico, R (Ed.) Proceedings of the 13th International Conference on Ground Penetrating Radar, University of Lecce, Lecce, Italy.

Utsi, E., 2013. Bringing up the bodies: high resolution and target definition using GPR. In: Proceedings of the Progress in Electromagnetic Research Symposium (PIERS 2013), Stockholm, Sweden, August 2013.

Utsi, E., Birtwisle, A., Cook, J., 2008. Detection of subsurface reflective cracking using GPR. In: Forde, M. (Ed.) Proceedings of the 12th International Structural Repair Conference, ECS Publications, Edinburgh, 2008.

Chapter 12

Examples of Practical Problems

Chapter Outline

All of the case studies in this chapter are based on real enquiries for potential projects and requests for technical advice. Once you have read the brief project description, try to answer the questions before reading the solutions below, if necessary referring to previous chapters. Varied as the projects are, this is the opportunity to try out your understanding of the basic principles of GPR detection.

CASE STUDY 1: THE POTENTIAL GAS EXPLOSION PROBLEM (REINFORCED CONCRETE)

Following a series of accidental gas explosions which resulted in loss of life because of buildings collapsing, structural engineers were legally obliged to assess a series of modular built (concrete slab) flats erected in the 1960s for their ability to withstand a gas explosion. The drawings and associated records still existed. However, all of the records had been adapted in the

Ground Penetrating Radar. DOI: http://dx.doi.org/10.1016/B978-0-08-102216-0.00012-6

course of building another series of flats some years later so that there was significant doubt over the accuracy of the remaining records, particularly as regarded the reinforcement within the concrete slabs.

Reinforcement of the concrete had been carried out on site and it was thought that the specified uniform style and size of rebar had been used throughout the development. In order to evaluate the stability of the overall structures in the event of a gas explosion, the engineers required to know the pattern of reinforcement, the plan of which was known to vary from one flat to the next but which should conform to a limited number of predesigned patterns.

Investigation of an empty flat revealed the following information:

● The reinforcement in the concrete slabs (ceiling/floor) used in the modular build consisted of a close knit metallic mesh at the top of each slab with reinforcement bars beneath.
● All ceilings had been covered in a render which contained significant amounts of asbestos.
● Electromagnetic methods of locating the position of the rebars could not be used because of the mesh above and the render below.
● Exposing a number of rebars proved to be expensive in time and money, partly because of the safety measures associated with the removal of asbestos.
● The rebars were not of uniform manufacture as had been expected but varied in size and shape.

It was therefore decided that a sample of flats, chosen by the structural engineers, would be investigated using another geophysical method, possibly GPR. Plans of the location (position and depth) of each rebar would be required for the slabs investigated. The engineers also wished to know if it was possible to estimate the size of the rebars. A random sample of the radar results would be verified by exposing the rebars and comparing the results of this investigation with the reported GPR data.

Question 1: What difference does the first bullet point above make?
Question 2: What other information would you want to know before attempting to locate the rebars?
Question 3: What frequency of antenna would you use and why? Given that you have access to 500 MHz, 900 MHz, 1.5 GHz, and 4 GHz, what critical calculation would you need to make in advance of survey? Make the calculation for each of these antennas.
Question 4: What should your first step on the survey be?

Case Study 1: Some Solutions

Question 1: Metal of any sort will be a strongly reflective target since metal is an electrical conductor. If we survey from the floor downward, it may be difficult to detect the rebars because the metal mesh above will reflect a significant proportion of the signal. This will inevitably reduce the signal reaching the rebars. The survey will have to be done upward into the concrete slabs forming the ceilings so that the first set of targets will be the rebars rather than the mesh.

Question 2: Are the ceilings smooth or has the render used been sculpted, ground coupling of the radar being the potential issue? What spacing was found (1) in the mesh and (2) between the rebars which were exposed? At what depth above the ceiling surface were the rebars exposed? This information is vital in order to ensure that the wavelength of the radar will penetrate between the metal, defining the position of each rebar. Strictly speaking the mesh information is probably redundant because of the potential loss in the signal before it reaches the rebars. The rebar spacing is essential since a high-frequency antenna will be required. The depth information is also helpful in determining which frequency of antenna can be used. There is little point in using so short a wavelength that the targets cannot be reached. It might also be wise to check if there are known damp problems in any of the investigation areas and also what the condition of the floors of the flats to be investigated is, i.e., would this impede movement of the GPR operator?

Question 3: The wavelength is the critical factor, both for target definition of the position of the rebars, the (low) possibility of being able to measure their size, and also to make sure that they will penetrate sufficiently deeply. As the investigation concerns internal ceilings, it is reasonable to assume (at least until told otherwise) a typically dry velocity of 0.1 m/ns although this will have to be calibrated during the actual survey.

Wavelength, $\lambda = v/f$, where v is the transmission velocity and f the frequency emitted (in GHz).

For the 500 MHz antenna, $\lambda = 0.1/0.5$ or 0.2 m (200 mm). Although this antenna would give plenty of probing depth, it is unlikely to be of use in terms of either target definition or passing between the metal reinforcement. Of course, its size and weight would probably also make it unsuitable for a ceiling survey.

For the 900 MHz antenna, $\lambda = 0.1/0.9$ or 0.11 m (111 mm).
For the 1.5 GHz antenna, $\lambda = 0.1/1.5$ or 0.067 m (67 mm). This is much more promising.
For the 4 GHz antenna, $\lambda = 0.1/4$ or 0.025 m (25 mm). This, too, is a possible choice.

Question 4: The previous question has given a range of possible antennas. The only missing link is their actual performance within the concrete. Concrete is a variable material which can be quite lossy (depending on the ion and moisture content). It therefore makes sense to arrange a trial in advance of survey and compare the outputs of the 1.5 and 4 GHz antennas, possibly also the 900 MHz antenna. Given the requirement to try and measure the size of the rebars, the highest possible frequency is the one we would wish to choose, subject to their being no significant problem of attenuation.

Results

The survey was carried out using a 4 GHz antenna along each of the four sides of the ceiling slabs. The sampling interval applied between readings was 8 mm, the lowest possible, for maximum definition. The probing depth required was less than 0.25 m so the radar was set to a time sweep of 5 ns. From this the depth and position of all rebars within each concrete slab was selected on the GPR data, using the first zero crossing of each hyperbola. The output from the processing software was then used to plot this information so that the engineers were able to relate the patterning to the type of slab used and, from this, determine the overall reinforcement patterning of each building investigated.

The ground truthing exercise confirmed the accuracy of the depths and locations after two minor adjustments, both of which confirmed the accuracy of the radar readings. The first adjustment needed was to take out the depth of render since the breakout of the reinforcement bars could only happen once the material containing asbestos had been removed. It was not necessary to remove any coating before placing the radar antenna on the ceiling and this depth was therefore included in the GPR data. The second apparent difference was the appearance of a series of very small strong amplitude signals closer to the ceiling surface than the rebars themselves. These were expected to be metallic because of the strong signal amplitude but were not found in the breakout until it was realized that the small metallic ties attached to the rebars were being removed along with the render.

As discussed in Chapter 3, More Fundamentals of GPR Operation, even with such a high-frequency antenna, it is very difficult to estimate the size of a small diameter object such as a rebar. The situation was complicated by the fact that many of the rebars were of different shapes, e.g., flanged, twisted, as well as different diameters. The solution used is beyond the scope of this introductory book but was based on signal amplitudes using the antennas in two different polarizations. Although this proved effective up to a point, it was far from foolproof. From the perspective of a GPR antenna, there is no great difference between a small flanged rebar and a simple one with a larger diameter since the metal influences the same cross-sectional area. Fig. 12.1 shows typical data from the survey. The information derived from the survey allowed the engineers to evaluate the stability of the buildings in the event of a gas explosion.

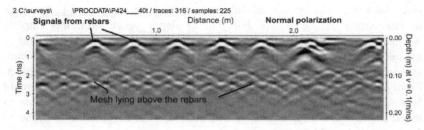

FIGURE 12.1 Definition of the position and depth of reinforcement bars in a ceiling (Utsi and Utsi, 2004).

CASE STUDY 2: DETECTION TO A DEPTH OF 5 M IN REINFORCED CONCRETE

The traffic authority in charge of a large Asian city was concerned about the development of sinkholes close to a popular shopping area within the city. A plan was drawn up to map an area using GPR with the intention of detecting any developing sinkholes and also any utilities present. Utility detection was required to a maximum depth of 1−1.5 m, sinkholes to a maximum depth of 5 m. The only ground information available was that the surface was reinforced concrete. There was no definite information on the spacing of the rebars but this was estimated at 125 mm. An e-mail request was received, asking which antenna would be the most appropriate for the proposed tasks.

> Question 1: In an ideal situation (i.e., assuming no attenuation), what frequency of antenna could you use for 5 m of depth penetration? Is this antenna suitable for the proposed project?
> Question 2: What is the lowest frequency of antenna which could be used to penetrate the reinforced concrete?
> Question 3: How deep could this antenna probe and what does this assume?

Case Study 2: Some Answers

The first piece of information to absorb from this request is that the targets include both potential sinkholes and utilities. Of these two, pipes and cables are likely to be the smaller targets. If we can locate the utilities, then we are likely to be able to detect the sinkholes, at least within the same depths. The full probing depth of 5 m is quite another matter. Given the presence of the reinforcement at near surface level, the problem is always going to be how low a frequency can we use without the rebars blocking the passage of the signal.

Question 1: Assuming no attenuation, the maximum probing depth we can hope for is 20 wavelengths. In reality, this is not likely to be achievable since both the conductivity and inherent lossiness of the soil will contribute to attenuation of the signal. For 5 m of depth penetration, this implies $\lambda = 0.25$ m simply to reach the full target depth. Since the wavelength is greater than the spacing between the rebars, 250 mm as opposed to 125 mm, any antenna capable of reaching 5 m depth would be useless for this project. From the point of view of the GPR, the close spacing of the reinforcement bars would appear to act like a metal sheet and it would not be possible to probe below the reinforcement. It will therefore not be possible to probe to 5 m depth below this concrete surface.

In practice, we would be very unlikely to get 20 wavelengths out of a GPR in this type of context so that, in order to reach the full depth, we would actually need to use a lower frequency antenna than this because of potential attenuation in the soil below the surface. Since sinkholes are often formed by the action of water in the subsurface, it would probably also be wise to allow for the effect of some moisture content. This is, however, academic because, if this particular antenna cannot probe beneath the concrete, there is no possibility of a lower frequency one with an even longer wavelength doing so.

Question 2: The lowest frequency of antenna which could be used on this concrete is one with a wavelength equal to or less than the spacing between the rebars, i.e., 125 mm. If we assume that any water content has drained and the remaining deposits are dry (something which would have to be checked), then the frequency in GHz = $v/\lambda = 0.1/0.125$ or 0.8 (800 MHz).

Question 3: Assuming no attenuation (which is very unlikely), the maximum depth of 20 wavelengths would be 2.5 m. This also assumes dry conditions. With a lower transmission speed, the wavelength would be shorter and the probing depth less. Taking attenuation into account will also restrict the probing depth. We might be lucky to reach the first requirement of $1-1.5$ m depth.

Results

The client was advised of the restrictions that made 5 m depth penetration impossible. Further soil analysis was sought before the utilities were mapped.

CASE STUDY 3: MOSAIC CONSTRUCTION AND GRAVES INSIDE A HISTORIC CHURCH

As part of a restoration program, a survey of a 13th century mosaic composed of many semiprecious stones and some glass, of varied size and shape, set in a marble background was commissioned. The survey was required to investigate the structure and condition of the mosaic in advance of cleaning and restoration and to detect any earlier historical features which might be concealed beneath. Subsurface soils were known to be free-draining and dry. Previous archeological investigation had shown that any building remains extended to no more than 2 m in depth.

Question 1: How would you decide what frequency of antenna to use?
Question 2: What factors would determine your survey strategy?
Question 3: How would you determine probing depth?
Question 4: How would you calibrate the transmission velocity?

Case Study 3: Some Solutions

Question 1: This is a classic multifrequency requirement. The mosaic is composed of small elements close to the surface which will require a high-frequency antenna to define their depths. The possible buried archeological material, being at greater depth, requires a longer wavelength and therefore a medium to low-frequency antenna. The actual frequencies used could be determined by calculation if we had sufficient information about the interior of the mosaic. As we do not, the wisest course is to carry out a presurvey test to establish the most suitable frequencies.

Question 2: Even large mosaics cover a relatively small area in terms of GPR surveys. This means that dense data sets should at least be considered for the spacing between adjacent survey lines. The high-frequency antenna should be set to its minimum sampling interval along the line of travel of the radar. The targets for the lower frequency antenna are potentially larger (building remains) and the sampling interval is therefore less critical. Total coverage of the area is obviously required to obtain the information requested.

Question 3: Although the information re probing depth may be accurate, it would be foolhardy to restrict the depth setting to 2 m in case this previously unexplored area either contained some deeper features or some unexpected moisture; 50−60 ns would be an acceptable starting point. Alternatively, a small number of random lines could be viewed in advance of survey to ensure that signal returns were not disappearing below the level of probing depth selected. The high-frequency antenna should be set to its maximum depth of penetration.

GPRs with preset depths will normally have a deeper setting than would be required here. This could impair the visibility of the data in real time since the full depth setting will be displayed on screen. However, the unused depth can be disregarded on postprocessing. The only potential disadvantage is that all radars sample in the vertical direction also and there will therefore be some loss of definition due to spreading the samples over a greater depth than is required. Under these circumstances, it is important to fit the depth setting to the probing requirement as closely as possible.

Question 4: Curve fitting is an acceptable answer provided that the data contain sufficient numbers of clearly defined hyperbolas. This is a reasonable expectation, given the composition of the mosaic from discrete, small elements. The ground beneath the mosaic should be independently calibrated from the mosaic itself in case the transmission velocity is different. It is a reasonable assumption that someone would have noticed had the mosaic been retaining water. The same may or may not be true of the deeper subsurface although the reason(s) for restoration might give a clue.

Results

Three antennas were trialed in advance of the survey: 4 GHz, 1 GHz, and 400 MHz. The 4 GHz gave very good target definition of the near surface. The 400 MHz gave good depth penetration for the deeper subsurface. The 1 GHz results were surprisingly lossy in comparison with the 4 GHz but this corresponds well to the research results regarding the different responses of the electrical and magnetic loss components (see Chapter 6: Attenuation or Loss).

A 4 GHz GPR was used to map the mosaic and the two burials revealed to lie directly beneath it. This gave information on the structure of the mosaic, the nature and manner of burials, and also some information on grave goods. The radar was set to its full probing depth of 10 ns, equivalent to 0.5 m in the dry conditions. The sampling interval along the line of travel of the radar was 5 mm and the spacing between adjacent survey lines was 5 cm. This latter measurement complied with the English Heritage guidance of the time (English Heritage, 2008). Although reducing this spacing in order to maximize the data collection was considered, the time slot and resources available for survey did not allow this to be applied to the full area. Velocity calibration made use of an existing, very small borehole through the mosaic and confirmed both the expected dry conditions and a transmission velocity of 0.1 m/ns. The voids associated with the two burials in the area lay directly beneath the mosaic. Since this meant that they were at a relatively shallow depth, it was possible to gain information on the contents of the graves using the 4 GHz antenna (Utsi, 2006).

A 400 MHz antenna was used to map the same area in order to examine any extant remains from an earlier (11th century) phase of the building. The sampling interval used was 5 cm and lines were spaced 25 cm apart which constitutes closer spacing than that recommended at the time. The probing depth of 60 ns confirmed that the extant remains were contained within 2 m of the subsurface. Velocity calibration, by curve fitting, confirmed the previous reading of 0.1 m/ns within the areas of soil and building remains.

The 400 MHz data revealed a number of buried structures, some of which relate to an earlier phase of the building. Comparison of the two data sets proved extremely helpful in interpreting the data from the two burials, notably in respect of the relative air gaps and the possible use of Fuller's Earth in the southern burial (at the lower end of Fig. 12.2). The success of the project led on to a more extensive program of GPR investigations of other areas of the building.

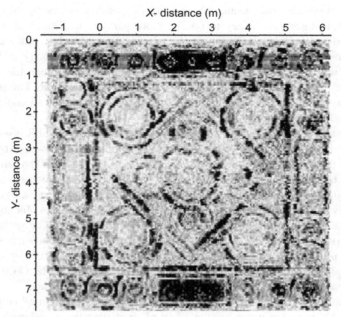

FIGURE 12.2 Time slice from the 4 GHz mosaic survey showing the lowest level of the mosaic elements and the top of two graves (Utsi, 2006).

CASE STUDY 4: A WETLAND ARCHEOLOGICAL SITE

In the late 19th century, during the construction of a boundary wall, a range of archeological artifacts were unearthed from the waterlogged peat in which they had lain since the late Bronze Age. These included a 4-foot high wooden figure, bog butter, and flints. Due to a shortage of sites on which housing could be built, planning permission had now been applied for. It was proposed to build houses in an adjacent large peat field lying between existing housing and the area of the boundary wall. An archeological GPR survey was commissioned in order to assess the archeological potential of the site before a decision was taken on the planning permission application. A previous investigation of the site using canes pushed into the ground revealed an undulating buried land surface of 0.1–3.5 m (Pollard, 1993).

Question 1: What survey strategy would be appropriate for this project?
Question 2: How will you decide on the frequency of GPR to be used?
Question 3: What impact will this have on target detection?
Question 4: How can transmission velocity be calibrated?

Case Study 4: Some Solutions

Question 1: The area is described as "large." Depending on the actual dimensions, it is worth considering a stratified sampling strategy, i.e., covering the full site at widely spaced intervals and then, depending on the results, applying a more intensive survey to one or more parts of it. This, and the time and funding available, will dictate the transect spacings. The probing depth needs to take account of the much slower transmission velocity associated with a waterlogged organic soil. The sampling interval along the line of travel of the radar will reflect the use of a low-frequency GPR. The possible use of a vehicle for towing purposes and ease of site access should also be discussed in advance of the survey.

Question 2: Since the site is waterlogged, a low-frequency GPR is essential. A depth of 3.5 m at a velocity of 0.035 m/ns implies a time sweep (two-way probing depth in nanoseconds time) of $t = 2d/v = 7/0.035$ or 200 ns. It would be wise to leave a margin in excess of this amount in case the denser sampling from using the radar reveals a deeper area not found by inserting canes at regular intervals. A radar with a set probing depth of 90 ns or less, for example, would be useless in this environment. The wavelength needs to be greater than 1/20 of (say) 4 m, i.e., 0.2 m although the exact value will depend on how lossy the wet peat environment is. At a velocity of 0.035 m/ns, $\lambda = 0.035/f$, so f has to be equal to or lower than $0.035/0.2 = 0.175$ GHz or 175 MHz. On the basis of that calculation, this would imply an antenna of 100 MHz or less.

Question 3: It would not be realistic to expect the detection of very small targets. Although the water has the effect of shortening the wavelength, it will still be relatively large and so any targets detected will also be relatively large. Individual flints, for example, would be invisible. Another 4-foot high figure might be large enough to be detected, depending on its orientation relative to the path of the radar and its water content relative to the surrounding organic soil. Whether or not such an object were detected would depend primarily on how great the electromagnetic difference was between the wood and its surrounding wet peat after burial for many centuries. The difference would be most noticeable if the peat and the wood had different water content. In a fully waterlogged context, this is possible but unlikely for two reasons. Firstly, full waterlogging implies a minimal difference between the electromagnetic response of the artifact and that of the surround peat. Both are composed of organic material with a high freshwater content. There needs to be a difference in electromagnetic response for the archeological material to be detectable. Given the high relative permittivity of water, this almost certainly means a difference in water content between the artifact and the peat around it. Secondly, if there is differential waterlogging, the artifact is likely to deteriorate since it is the water content that has preserved it (anaerobic conditions). The degradation of the artifact makes it less detectable and potentially renders it undetectable.

Question 4: Peat basins traditionally have flat surfaces in the center so that, if a radar with separable antennas is used, a WARR calibration can be used. Failing that coring at a predetermined position could be used. In practice, this would be at a number of predetermined positions rather than a single one. As a last resort, curve fitting might be used.

Results

A 50 MHz radar with separable antennas was used to map the full Peat Moss at 10 m intervals. This revealed anomalies centered around one particular area, the deepest part which was then surveyed at 2 m intervals. WARR calibration was used in all parts of the site to give a uniform transmission velocity of 0.035 m/ns. A small trial excavation was carried out after the survey. This confirmed the existence of human activity in the form of a worked wooden peg adjacent to a clay platform within the peat. The paleo-botanical evidence indicated that, at the time the platforms were constructed, the peat would have been dry. Fig. 12.3 shows a two-dimensional profile of the center of the survey area and Fig. 12.4 one of the time slices illustrating the position of the platforms relative to the surrounding granite.

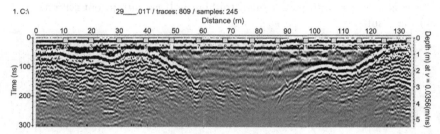

FIGURE 12.3 Radargram across the Peat Moss showing the outline of the underlying granite and some of the anomalies within the peat.

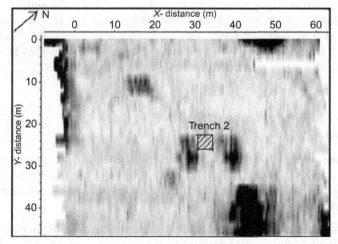

FIGURE 12.4 Time slice showing the surrounding granite on the outside of the Peat Moss, four platforms within the peat (not all at the same depth), and the position of the excavation trench (Utsi, 2004).

If a line is drawn between the center of the granite headland visible at the foot of the time slice (and on the right-hand side of Fig. 12.3) and the center of the furthest away platform, this aligns with the East-West axis of the site (see also Clarke et al., 1999; Utsi, 2004).

CASE STUDY 5: CARBONATE CONCRETIONS IN A SALT MARSH

As a GPR surveyor, you are asked to assist a research project in the detection of carbonate concretions. The expected depth for these is anything between 1 and 3 m. Target size is small, of the order of 10−20 cm. The only other restriction is that the GPR used for detection must be portable because of the terrain, a salt marsh, described as wet and clayey. As it is a research project, ground truthing will be carried out on a sample of the results.

Question 1: What frequency of antenna would you use on this project?
Question 2: How would you make this decision?
Question 3: How would you calibrate the transmission velocity?
Question 4: Would a trial survey be helpful?

Case Study 5: Some Answers

There are two, possibly three words in this project description which answer any or all GPR questions. Depending on the accuracy of the initial description, it might be appropriate to make a short trial but only with the caveat clearly expressed to anyone funding the project that this is almost certainly not a viable project for GPR. The relevant words are "salt," "wet," and "clayey." If in doubt or if you attempted to answer the questions in the format presented above, re-read Chapter 6, Attenuation or Loss.

CASE STUDY 6: A DELAMINATING FLOOR IN A SUPERMARKET

Tiles were working loose from the floor in a recently built supermarket. In order to prevent further accidents, a trial using GPR was arranged on a mock-up of the floor to prove that radar could detect delamination of the tiles from the floor.

Question 1: In radar terms what is the principal problem (apart from Health and safety considerations)?
Question 2: What are the implications of your answer to question 1?

Case Study 6: Some Solutions

Question 1: The tiles are not very thick and the grout that holds them on the floor is thinner. Any voids developing are therefore going to be very small targets.

Question 2: The first implication is that a high-frequency GPR is required, the higher the frequency (the shorter the wavelength), the better since it is desirable to detect the development of an air gap before the tile actually lifts off. The second is that since it is a question of very small targets very close to the ground surface, in theory at least this should be a suitable task for a GPR.

Results

A 6 GHz antenna was used to map the floor. A time slice, extracted at a depth of 32 mm is shown in Fig. 12.5. The survey lines are relatively widely spaced (1 per floor tile) but the sampling interval along the line of travel has been kept low (c. 5 mm). The very strong signals (in black) indicate where the grouting is inadequate and there are voids present. This is a relatively rapid solution to determining how much re-grouting requires to be done.

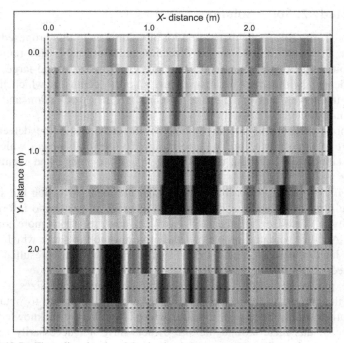

FIGURE 12.5 Time slice showing delamination below certain floor tiles.

Note that the more common application of detecting delamination in or below the wearing course of a road or bridge using GPR follows exactly the same logic and principles although the structural context may be more complicated, particularly for a bridge.

CASE STUDY 7: DETECTION OF UXO IN REDEPOSITED EARTH

A port city has been importing soil in order to protect its shoreline from being worn away by seasonal storms. In the course of reconstructing and extending one particular industrial site, the city authorities have discovered that the soil being used to extend and protect the location contains unexploded ordnance, not all of which has metallic content. This site already contains at least one factory area and the site needs to be safe so that it can be properly developed and used for additional manufacturing plant. The planned foundations for these buildings extend several meters underground.

Question 1: What, in GPR terms, are the principal problems of using GPR to detect UXO and clear the site?
Question 2: What would you do before taking on such a survey?

Case Study 7: Some Solutions

Question 1. In GPR terms there are a lot of unknowns in this proposed investigation. Given that it concerns the detection and presumably safe removal of dangerous targets, this is a very serious matter. The size of the targets, their likely density within the soil matrix, and also the depths at which they lie are all unknowns that could make antenna frequency selection and hence detection difficult.

Although there are specialist GPRs available suited to the detection of this sort of target, clearance will be required to several meters which may not be compatible with the optimal wavelengths to image and identify the targets.

It will probably be possible to determine the depth to which the site has been built up in advance of development. There is currently no information on the type of soil or whether soil has been imported from more than one source. Another potential problem is the proximity to the shore which should not yet have affected the newly deposited soil but will have resulted in salt contained within the original soil.

Unexploded ordnance can cover a multitude of types and sizes of device and further information is important. It may not be possible to obtain this information simply because the soil provider is unlikely to acknowledge the presence of this material and the site owners have not necessarily seen the full range of potential devices.

Question 2. The first requirement of this project is to establish whether or not it is going to be possible to use GPR. It is therefore essential to clarify as much of the missing information as possible. It is critical to have soil information, i.e., whether there is more than one source of soil, what the electromagnetic response of the soils, including that of the original soil if this lies within range of the foundations, are and also confirmation of the depth to which the site conservation measures have been carried out. There is a significant risk that this information, combined with whatever information is available about the size and shape of the devices, is likely to rule out detection by GPR on the basis of incompatibility between the size of target to be detected and the depth required to be probed. There is also a real risk of the various soils being unsuited to survey by GPR. There is no point in taking the GPR project any further if these two risks are not resolved.

Depending on the amount of experience that you have in the field of GPR detection of unexploded ordnance, you would be wise to consult a specialist, particularly one with knowledge of the recent history of the area from which the imported soils have been sourced.

CASE STUDY 8: BRIDGEDECK CONSTRUCTION

An engineering firm wishes to commission a GPR survey of a large bridge which carries a major road. The bridge construction is a wearing layer of asphalt over reinforced concrete. The bridge is developing problems as rainwater permeates into cracks in the asphalt layer which risks causing delamination of the asphalt layer. Due to the volume of traffic carried by the road, it will not be possible to close the bridge during the survey. No information is available on the spacing of the reinforcement in the concrete layers. The engineers wish to know the depth of each construction layer, the condition of the reinforced concrete, and the location of any delamination between the asphalt and the underlying concrete. If the survey is successful, there may be a requirement to investigate tenon joints.

Question 1. What are the GPR implications of the need to comply with the road requirements of the survey?
Question 2. What frequency or frequencies of antenna would be suitable for this investigation?

Case Study 8: Some Solutions

Question 1. Since the road cannot be closed this survey requires a vehicle. It may also require nighttime working. If using a vehicle, then the use of horn antennas should be considered since for working at speed, it will not be possible to ground couple the antennas. The antennas will have to be securely fixed to the vehicle and consideration will have to be given to how an accurate distance measurement is to be achieved.

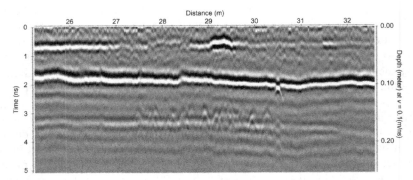

FIGURE 12.6 Section of a bridge survey using a 4 GHz horn antenna, showing the strati-graphic layers of construction as well as areas of delamination below the wearing course. The reinforcement faintly visible is part of a bridge pillar (Utsi et al., 2014).

Site coverage is also an issue. The vehicle will need to travel in lane during the survey and, in order to achieve reasonable site coverage, it is likely that the antennas will have to be moved from one position to another in order to gain at least the information from the three basic positions for each lane of the carriageway (see Chapter).

Question 2. This is a classic multichannel, multifrequency survey. Delamination is best detected with a very high-frequency antenna of between 2 and 6 GHz although there is evidence to suggest that, for a survey at speed using a horn antenna, the frequency ought to be 3 GHz or above (Utsi et al., 2014). The stratigraphic information can be obtained from either a 1.5 GHz antenna or possibly a 400/500 MHz antenna, depending on the total depths of the road and bridge. It is unlikely that the very high-frequency antenna will be able to define the reinforcement bars with great clarity given the depth constraints of this antenna and the likelihood of strong reflections from areas of delamination between the asphalt and the concrete above.

Results

Fig. 12.6 shows an example of typical output from this sort of investigation. Although these GPR images would be used for illustration in the GPR report, the information requested by the engineers would normally be provided on scale drawings, using the output from the GPR data interpretation software package.

REFERENCES

Clarke, C.M., Utsi, E., Utsi, V., 1999. Ground penetrating radar investigations at North Ballachulish Moss, Highland, Scotland. Archaeol. Prospect. 6, 107–121.

Pollard, A., 1993. Archaeological Assessment of Ballachulish Peat Moss, Highland Region. Glasgow University Archaeological Research Division on behalf of Highland Regional Council (GUARD report 136, Glasgow).

Utsi, E., 2004. Ground-penetrating radar time-slices from North Ballachulish Moss. Archaeol. Prospect. 11, 65–75.

Utsi, E., 2006. Improving definition: GPR investigations at Westminster Abbey. In: Daniels, J., Chen, C.C. (Eds.), Proceedings of the 11th International Conference on Ground Penetrating Radar. Columbus, OH.

Utsi, V., Utsi, E., 2004. Measurement of reinforcement bar depths and diameters in concrete. In: Slob, E., Yarovoy, A., Rhebergen, J. (Eds.), Proceedings of the 10th International Conference on Ground Penetrating Radar. Delft, The Netherlands, pp. 659–662.

Utsi, E., Birtwisle, A., Utsi, V., 2014. GPR analysis of Bridgedeck construction and the detection of delamination. In: Forde, M. (Ed.), Proceedings of the 15th International Structural Faults and Repair Conference. Imperial College, London.

English Heritage, 2008. Geophysical Survey in Archaeological Field Evaluation. English Heritage Publishing, Swindon, UK.

Chapter 13

Multichannel and Single Channel Systems

Chapter Outline

There are a wide variety of GPR systems available on the market and what suits one person or application may very well not suit another. The purpose of this chapter is not to assess any or all of the individual systems available but to draw attention to some of the features which might be useful to know before deciding which system is the most appropriate for the investigation and the site. Many GPRs are multipurpose investigation tools but there are also a number intended primarily for specialized investigations. Whether this makes them suitable for other GPR surveys or not is something which may need consideration. As outlined earlier in this book, it is important to assess whether the GPR equipment intended for the survey has the right characteristics for the survey or not. If not, then the choice remains either to hire in the correct equipment or to pass the survey on to someone who has a more suitable GPR.

There has been an unfortunate tendency among some GPR users, particularly those with limited practical experience, to use the GPR equipment readily available without fully assessing its suitability for their specific project. Unfortunate results of this which have come to the author's attention include attempting to use a 400 MHz antenna on a reinforced road where the rebar spacing was too small, the failure to verify the location of certain graves of historic interest, the failure to locate a large sewer pipe buried in damp open ground, and failing to find an electricity cable in the wet soil of a redevelopment site. Caught in this type of situation, there is a human tendency to blame GPR technology instead of recognizing that the radar used in each case was inappropriate for the investigation. From this, comes some of the urban mythology such as the idea that GPR cannot be used in water. Following the guidelines given in earlier chapters of this book should enable a GPR surveyor to avoid such unsatisfactory outcomes.

Ground Penetrating Radar. DOI: http://dx.doi.org/10.1016/B978-0-08-102216-0.00013-8

As discussed in earlier chapters, it is important that the correct frequency/wavelength is selected for the survey. After that, the most important consideration is suitability of the GPR for the survey site. Even within the same GPR application, survey sites vary enormously in the type of terrain to be covered, potential surface obstacles, soil type, and even the size and range of targets. The equipment suitable for one particular site may not be equally suitable for all similar investigations. For example, many utility surveys are either on urban sites or development sites but they can also be carried out over open fields or footpaths. Archeological surveys are often in open countryside but can equally well be on development sites or indoors within standing buildings. Whatever the site surface, it is important to have a system which will remain ground coupled during the survey (cf Chapter 3: Velocity Calibration). The first decision is to make sure that a trolley or towing skid appropriate to operation on the site surface is selected. For vehicle mounted or towed systems, the antennas need to be placed within one-eighth of a wavelength of the ground surface, and there are safety issues to be considered in the way in which the antennas are attached, particularly if they are to be used in the vicinity of other vehicles and/or the mounting position will have to be changed in order to obtain the requisite site coverage.

The next decision is between single channel and multichannel systems. Depending on the frequency of antenna to be used, there may not be options for multiple channels. Typically many of the low-frequency systems (100 MHz and below) are single channel systems, if only because of the size of the antennas. Otherwise, the site dimensions may make it more or less suitable for coverage by a multichannel array. One consideration which might favor a single channel system over a multichannel, even for relatively large sites, is the number and regularity of surface obstacles. As an example, when carrying out the survey of the nave of Westminster Abbey, the author opted for a single channel system over a multichannel in order to gather as much information as possible around the pillars in the Nave. An array of antennas would have left unnecessary gaps in the data around each pillar. The alternative would have been to use a multichannel for the area between the pillars and a single channel around the pillars. As the pillars were large enough to affect data collection along several adjacent survey lines, a single channel system was preferred as this can be worked around the pillars more easily and with fewer data gaps than a multichannel.

SINGLE CHANNEL SYSTEMS

Single channel systems are the basic workhorses of the GPR world. A single channel consists of a control system with one transmitter and one receiver of the same frequency. The antennas are usually, but not always, housed within a single antenna casing. Some single channel systems have separable antennas.

Single channel systems often have a choice of antennas of different frequencies. This means at least one alternative antenna, sometimes a full range of antennas of different frequencies. As the controller can only operate one transducer pair at a time, the antennas cannot be used simultaneously. A transducer pair is one transmitter and one receiver of the same frequency. If the investigation requires data from more than one frequency, different sets of antennas will have to be used one at a time.

Some single channel systems are designed specifically for particular applications, e.g., utility detection. Most, however, are designed for general-purpose use on whichever application suits their range of antennas. Most come equipped with a means of mobility, typically a trolley, but many can also be hand towed where the site surface is more suited to this.

In terms of the way in which the radar is set up and the way in which different survey parameters are chosen, the individual systems vary widely between manufacturers, all of whom have their individual preferences. All systems record their data usually in the manufacturer's preferred format. It is still possible to find systems which record data in pictorial form although these are becoming fewer. This is not an ideal way in which to record survey data as it means that the data cannot be interrogated after the survey, should the need arise. It is better to have a radar from which the data can be downloaded and processed, even if the surveyor's initial intention is to work from what is visible on-screen. This is particularly important when working in utility detection where it is relatively easy for pipes and cables to be obscured by other subsurface material. Systems which use processing presets record processed data, applying the processes selected for the survey automatically to the data. Other systems retain the raw data, independently of the on-screen settings applied during the survey. The data from either type of GPR can be postprocessed and it is a matter of personal choice as to which is preferable.

In general, beyond the choice of the number of antennas, single channel systems have the same or very similar data presentation options as multichannel radars. The depth setting may be selectable or it may be fixed. If selectable, the choice may be in nanoseconds time or categorized as "shallow"/"deep," for example. There is usually provision for depth settings in meters (or feet) by selection of a value for either the transmission velocity or the relative permittivity, Er, as part of the survey parameter setup. There is always a choice as to the sampling interval along the line of travel of the radar, either in terms of a number of readings per meter (or per foot) or in terms of a sampling interval of so many centimeters (or inches). In addition there is usually the possibility of operating the radar continuously although this would normally be used only for setting up the radar, not for surveying. Most systems have provision for at least one marker which can be applied to the data during survey, some have options for more than one.

Settings for viewing the data will also vary with the radar used. Some GPRs apply gain to compensate for attenuation automatically, others give the operator choices of gain to apply to the data. Some systems have all parameters set up in advance of survey, including gain compensation, others allow the user to change the value of gain during survey. The operating software may include other features such as monitoring for data quality, calibration of depth, and removal of Tzero.

The most commonly used frequencies for single channel GPRs are the mid-range frequencies of 250 MHz, 400/500 MHz, 900 MHz/1 GHz, and 1.5 GHz. That said, the appropriate frequency depends on both the GPR application and the particular circumstances of the site. The very low-frequency snake antennas, for example, whose frequencies lie typically in the 10–80 MHz band are also single channel systems.

Even where multiple antennas are available, single channel systems use only one transducer pair at a time. If a second or subsequent frequency is used, it will be necessary to change the antennas. Single channel radars do not differ significantly from their multichannel counterparts in their use of built-in screens, laptops, notebooks, or tablets. These options are usually set by the manufacturer for both types of GPR.

Single channel systems are commonly used from trolleys which can carry the control systems and the antennas. For rougher surfaces, there is usually a hand towing solution and for investigations of ceilings or walls, the antennas are usually mounted on some form of handheld skid.

MULTICHANNEL GPRs

There are many different types of multichannel GPRs. The term multichannel covers any radar that uses more than one set of transmit and receive antennas at the same time. The usual aim of a multichannel is to cut down on the amount of time spent surveying either from using more than one set of antennas of the same frequency or from using several frequencies simultaneously.

The two main types are therefore multiple channels using the same frequency of antenna and multichannel, multifrequency. The former are intended to ensure greater ground coverage each time a survey line is recorded. So, for example, a road investigation could usefully use three channels of 400 MHz to cover the center of the lane and the mid-near side and offside positions or more channels of the same frequency if greater coverage was required.

A similar requirement comes from the use of multichannel radars to cover large swathes of ground in archeological surveys. A good example of this is the recent detailed GPR survey work carried out in the vicinity of Stonehenge in the United Kingdom using a radar with 16–31 channel capacity (Gaffney et al., 2012). Similar technology has been applied in the field of detection of unexploded ordnance (UXO). One obvious use of GPR technology is to allow vehicles to travel along roads and tracks safely from one place to another. UXO detection sweeps are routinely done with multichannel radars mounted on the front of a vehicle. The precise configuration of frequencies in use varies depending on the manufacturer of the equipment but the number of channels used may be anything from 8 to 32. This depends in part on the frequencies used since the coverage is obviously more focused for higher frequencies of antenna, potentially requiring more antennas for the same area coverage.

The use of an array of antennas cuts down the survey time by covering a larger area with each pass of the radar. How many channels are used very much depends on the purpose of the survey and the amount of ground required to cover on each survey track. Unless the data processing can be automated, there is no reduction in the data processing time since the potential amount of data which can be generated is increased. Automated processing is usual only where the site conditions can be predetermined on the basis of experience. UXO detection is one obvious example where this is both possible and also necessary for the survey vehicle not to be destroyed while surveying.

Multiple frequency multichannel radars can also be used to cover wider tracts of ground for each sweep of the radar than would be possible with a single channel. However their main purpose is to use several frequencies of antenna at the same time. This is typical of road surveys, for example, where the combination of a lower frequency such as 400 MHz with a high frequency, such as 1.5 GHz, allows the surveyor to carry out examinations at depth at the same time as collecting more detailed data in the nearer subsurface. High-frequency combinations of antennas in the range of 1−6 GHz are often deployed for bridge deck investigations where detecting through reinforcement is an issue.

It depends on the manufacturer as to whether the antenna positions are predetermined or whether the gaps between adjacent antennas can be altered to suit the application. Mechanically the former tend to be more robust but the inability to fix antennas along different spacings does not always provide the optimum coverage for any given site.

When using a multichannel over an area (as opposed to along a number of discrete survey lines), care needs to be taken over the placing of the lines. If the survey direction is the same on each pass of the radar, there is no risk of confusion. If the data are collected in a zigzag or meandering fashion, i.e., by starting at alternate ends of the survey area then the order of the channels changes on each sweep and needs to be placed accordingly in the three-dimensional (3D) data. As an example, consider the case of an eight-channel radar surveying between line A and line B. All eight lines are parallel to each other and an equal distance apart. On survey line 1, the eight channels will represent the data in the same order as on the equipment, i.e., 1−8. On survey line 2, carried out in the reverse direction, channel 8 is the next in the sequence and channel 1 the last. Since this is the second set of survey lines, channel 8 will be the 9th survey line, channel 7 will be the 10th, and so on down to channel 2 which will be the 15th and channel 1 which will be the 16th. A similar logic applies to multifrequency antennas of the same frequency.

DUOS AND TRIOS

A relatively recent development in the field of multichannel radars is "duos." These are generally intended for the purpose of utility detection where it is recognized that it may be advantageous to use more than one frequency, not only because of the variation in target size and depth but also because it is recognized that urban ground conditions may vary rapidly even within the same site, e.g., either from a variation in water content or because of imported backfill of inconsistent quality since recycling of material is becoming increasingly common. The dilemma between achieving an adequate probing depth and also good target definition can be addressed by using data from a combination of frequencies. There are two possible duo designs, those which contain two separate antennas transmitting different frequencies and those which contain a single antenna but use digital filters to extract two different frequency bands. The former is to be preferred since it makes use of the peak energies of both antennas (cf Fig. 2.1).

Another recent addition to the market is a tripartite design which operates at three separate frequencies (250 MHz, 500 MHz, and 1 GHz). According to the manufacturer, this design comprises three separate antennas built in Russian doll fashion into a single antenna casing with separate internal screening to reduce potential cross channel interference. As such, this is a true multichannel with three peak energies at the three nominal frequencies.

For both duos and trios, there are critical design factors in the internal coupling within the antenna boxing, notably in any interaction between the screening and the antennas. As touched on earlier, the combination of antennas and screens can lead to unforeseen unexpected performance (effectively interference) unless the antennas are electrically separated. This is not something that a GPR user should attempt to alter but it is something to be aware of and another good reason to arrange a trial of the equipment, something which all good manufacturers and their agents generally encourage.

SWEPT FREQUENCY SYSTEMS

The radar operation described up to this point concerns impulse GPRs. Impulse GPRs emit pulses at regular intervals (dictated by the speed of the system controller and the sampling interval chosen) to a prescribed depth in nanoseconds, either preselected or selected by the GPR operator. The frequency at which the impulse radar operates is dictated by the frequency at which the peak energy is emitted. The frequency is effectively a constant (or range of constants) and the systems are said to operate in the time domain, i.e., it is the variation in signals with the depth in time that is measured.

All electromagnetic signals can be represented in time but they can also be represented in varying frequency rather than time. Swept frequency systems operate by transmitting parts of the pulse in sequence, thus sweeping through the full specified frequency range of the instrument, effectively detecting the variation in signal response with frequency. These systems can offer a wider range of frequencies and potentially also lower noise than traditional impulse systems. However, not all practical survey results have confirmed this (Linford et al., 2010; Leckebusch, 2011). Further design modifications resulting from these findings may have improved the equipment performance. The data output from swept frequency systems requires more specialized data processing than the impulse systems (e.g., Persico, 2014).

Although there are fewer of these systems in general use, they too come in multiple arrays and are generally used to cover a wide area with each survey line of the radar. Their ability to characterize sites by the variation in frequency neatly sidesteps the question of which is the optimal frequency of antenna that a GPR surveyor should use on site.

CURRENT AND FUTURE DEVELOPMENTS

GPR manufacturers have a very good understanding of the challenges which face their clients in field operation and continue to introduce improvements into equipment performance and also the coordination of their equipment with other geophysical techniques. Since utility detection is the largest single use of GPR, this area continues to receive a great deal of attention. Recent developments include the introduction, by at least one manufacturer, of radio detection sensors into GPR equipment in order to assist in highlighting the existence of cables by overlaying data from the detection sensor with that being generated by the GPR. The aim is to both reduce, where possible, the time necessary on site, and improve detection rates.

The increasing density of subsurface utilities has also led to the introduction of multifrequency, multichannel dense array systems intended to enable the construction of detailed 3D images of large underground areas through which it is hoped to improve detection of utilities. Software developments to accompany this feature real-time automatic detection of underground assets.

In more general terms, recent advances in the cost and design of high-speed samplers have allowed several companies to improve the signal-to-noise ratios of their equipment (essentially the inherent clarity of the data produced) by improving the stacking ratios of successive data samples, without impeding the speed of operation of the radar. Related to this process, real-time samplers are also under development or have been recently introduced by a number of manufacturers, again improving the internal stacking ratios from which the increased sensitivity should provide better depth performance.

Note that the developments referred to above are not to be confused with more controversial claims in improved performance. In each case, the manufacturer is attempting to improve the output from the radar without attempting to deny or ignore the well-known limitations of all radar performance such as attenuation of the transmitted signals in ionized soils or a lack of unlimited energy. All manufacturers of reputable GPR equipment find inappropriate technical claims irritating, if only because they may be left with the fallout from equipment which does not conform to expectation. Most manufacturers therefore provide technical information via their websites and social media such as YouTube. At least one major manufacturer is in the process of moving as much of their training material as possible into the public domain in this way which can only help in terms of educating both users and their clients into what is and is not possible.

All of which begs the question, faced with a number of highly competitive and competent manufacturers of GPR equipment, and with the risk of not always recognizing inappropriate equipment, what steps should the novice user take? The answer is relatively simple. Always ask for a demonstration of the equipment. Arrange to hire before you buy, making sure that you get proper instruction in the use of the equipment first. Background reading or viewing online should give you a good idea of what the GPR world as a whole considers acceptable. And, if you are looking for a service rather than training in the technique, look for professional accreditation (see Appendix B: Professional Association and Reputable Manufacturers).

REFERENCES

Gaffney, C., Gaffney, V., Neubauer, W., Baldwin, E., Chapman, H., Garwood, P., et al., 2012. The Stonehenge hidden landscapes project. Archaeol. Prospect. Vol. 19 (2), 147–155.

Leckebusch, J., 2011. Comparison of a stepped-frequency continuous wave and a pulsed GPR system. Archaeol. Prospect. 18 (1), 15–25.

Linford, N., Linford, P., Martin, L., Payne, A., 2010. Stepped frequency ground-penetrating radar survey with a multi-element array antenna: results from field application on archaeological sites. Archaeol. Prospect. 17 (3), 187–198.

Persico, R., 2014. Introduction to Ground Penetrating Radar: Inverse Scattering and Data Processing. John Wiley & Sons, Hoboken, NJ, pp. 32–47. ISBN 978-1-118-30500-3.

Chapter 14

Reports and Publication

Chapter Outline

A very important part of any GPR survey is reporting on the findings, preferably in a manner which can be easily understood by the client and providing the information required in the original brief. Where some or all of the information cannot be provided, it is important that the reasons for this are given clearly. Unfortunately the reasons given for a lack of success do not always hold true and we will examine some of these at the end of this chapter.

As with all geophysical surveys, the data from GPR surveys should be retained in an accessible archive whether or not it is postprocessed. It is good practice to back up the data in a suitable storage medium before beginning processing data on a separate computing device. Once the project has finished, the processed data can be added to the backup archive. This is not to say that all data should be kept in a current file or folder but it is important that it should be accessible in the event of a query both in the immediate aftermath of the survey and in the more distant future. For certain data sets, there are also third-party archives to consider.

The requirements of a GPR report are the same as those for any other geophysical survey technique. There needs to be clear statements of what is being reported on, who the report has been compiled by, the intended recipients of the report, and how the investigation was carried out, including the results and conclusions. There are a number of prescriptive lists on the contents of geophysical reports, notably in Historic England's (formerly English Heritage's) guidance to archeological geophysicists (English Heritage, 2008). In the field of utility detection, there are prescribed document flows from the initial method statement down to the final utility drawing (ASCE 2002; PAS 128, 2014).

Ground Penetrating Radar. DOI: http://dx.doi.org/10.1016/B978-0-08-102216-0.00014-X

REPORT TITLE AND INTRODUCTION

As with all reports, the GPR report should have a title, the name of the client for whom the work was done, the name(s) of the author(s), a date, and, where applicable, a reference number.

The report introduction should set out clearly the aims of the survey work including any relevant background information and the reasons why the survey has been commissioned. The person(s) or body commissioning the work should be mentioned in the introduction as should the site location.

METHODOLOGY

This is one of the critical elements of a GPR report. It should refer back to the aims and background of the introduction so that there is a clear link between the aims of the survey and the manner in which this was to be achieved. This link must take account of the capabilities of the GPR used and the realistic limitations of the method. It would not be sensible, for example, to select a radar which did not have the required depth capability, including allowing for moisture content where applicable. Equally using too low a frequency of GPR antenna to penetrate reinforced concrete would not be an acceptable survey strategy.

Depending on the nature of the survey targets, it is useful to explain the basis on which GPR will detect these. So, for example, voids, metal, and even water may be identifiable on the basis of signal amplitude since all of these represent major changes in electrical and magnetic properties from their surrounding environment. Pipes and ducts, the footings of buildings and archeological structures might be identifiable from patterning in the horizontal plane, i.e., from time slices. The reader should not be left with the impression that the radar can identify specific materials as though the radar were an optical device capable of "seeing" individual materials. Nor should they be left with the impression that the radar was capable of measuring a discrete property of the target material other than the electromagnetic responses at its interfaces. Since the radar effectively maps the changes in electrical and magnetic properties from one material to another, there will always be some uncertainty associated with target detection. The basis on which the results have been analyzed and categorized should be made clear in the methodology section.

The description of the fieldwork should include the date(s) on which this was carried out and the type of GPR equipment used, including the frequency(ies) of antennas used. The survey parameters selected by the survey team should be clearly stated and these should be on the right order of magnitude to make sure that all targets have a good chance of detection. Where a sampling strategy has been applied to the site, the reasons for this should be clearly explained so that the reader has a chance of understanding the likely risk of nondetection. Any steps taken to minimize the risk of nondetection as a result of this strategy should be spelled out.

It is usual to include a diagram, plan, or photograph of the survey area in order to illustrate the site layout. The position of survey reference lines (or grid), if used, should be recorded. If GPS is used with or without external reference points, then this should also be stated. Survey direction should be noted and related to how the targets are to be detected. Where appropriate, the justification for single direction survey should be made clear. For example, within a Christian church, burials are normally oriented East/West and it would therefore be reasonable to survey along a North/South axis in order to get the best possible target definition. However, there will be some areas in most churches where there simply is not sufficient room to survey along a North/South axis and it may be necessary, in order to maximize the number of pixels in the three-dimensional data, to survey along an East/West orientation. Where two orthogonal directions have been used, this should be stated. If the sampling parameters along the two axes differ, then the justification for this should be in this section of the report.

As discussed in Chapter 5, Velocity Calibration, velocity calibration is critical to accurate depth measurement in meters/centimeters or feet/inches. The method of calibration (or methods if more than one has been used) should be given as well as the reasons for selecting the method(s) chosen. The results of the calibration should also be reported, particularly if these vary across different sections of the survey site. There should also be a statement of which transmission velocity or velocities have been used in the analysis of the results presented in the report.

RESULTS

A summary of survey results is often presented toward the beginning of the report with a more detailed description further on, usually following the introduction and methodology. It is not always necessary to include the complete data set in the results section of the report. It is more usual to pick representative examples of the information being reported on. The amount of evidence that will be required depends on the type of survey and also the amount and type of information required by the person or company commissioning the survey.

Where data have been processed, there should be a list of which processing steps have been applied to the data and, if possible, a short note of why these particular processes were chosen. If markers have been used on the data, either to indicate the position of the survey lines or to indicate some other aboveground obstacle or material, the system used for marking should be explained. Any data presented as evidence in the report should display the markers.

Evidence used may be two-dimensional data, a combination of two-dimensional and three-dimensional data, or time slices extracted from the three-dimensional data set. Where three-dimensional data have been created either in postprocessing or automatically on-site, this should be cross-checked with the two-dimensional data set before coming to any conclusions on the nature of subsurface targets whether both data sets are presented as evidence or not. This is because this comparison can sometimes give more information about the source of the returned signals than if only one of the data sets is examined in any detail. It is not really appropriate, for example, to examine only the time slices, looking for patterns in the horizontal plane without cross checking to the two-dimensional data. Whatever evidence is used, the manner- and orientation in which the data are presented should be indicated.

For many GPR applications, it will be appropriate to include plans show- ing the position of detected targets. This is mandatory for utility surveys and the format in which these data are presented is prescribed. It is also a normal requirement of civil or structural engineering surveys. It is usually necessary to interpret the data in a manner which is suitable for the client without assuming any great knowledge of radar technology on their part. In these cir- cumstances it will be probably also be appropriate to produce one or more plans of the site interpretation. The time slices can sometimes be used for this purpose. This depends on the site coverage and the clarity of the time slices. It is obviously easier to use time slices in this way if there is total site coverage than if strategic sampling was used.

It is important that interpretations of the data are backed up either by appropriate evidence in the GPR data set or by external evidence. The type of evidence will depend on the survey application but includes evidence from coring, test excavation, aboveground pipe labeling, connection to iden- tifiable utility sources such as electricity substations, and, in certain circum- stances, may include historical documentation.

It is also important to avoid overinterpretation of the data. For example, a series of anomalies of roughly the same signal amplitude arranged along three sides of a potential rectangle could be interpreted as the remains of the footings of a former building or possibly part of a former utility network. However, if these anomalies do not join up in horizontal view, this would not be a reasonable interpretation. The similarity in signal amplitude might indicate the same materials in use but it might not. The signal amplitude reflects the magnitude of the change, not the existence of any particular material. It is possible for this to be different materials for each point of the anomalous data. In this case the apparent placement of the signals could be completely misleading.

Another well-known example of overinterpretation involves a GPR sur- vey carried out to identify archeological features in an urban landscape. The survey results appeared to show a large number of features in one particular location. However, on excavation, this turned out to be a deeply buried raft of reinforced concrete and not the expected historical building remains (McCann, 1995).

It is important to be sure of the depths when interpreting the results of the survey. Wherever possible, data from uneven or sloping ground should be topographically corrected before any attempt is made to interpret the sub-surface features. Although it might be possible to identify a flat floor below a mound on the basis of the inverted dip (see also Chapter 10: Three-Dimensional Data), more complicated structures would not be so easy to interpret and it would be difficult in any case to extract time slices which demonstrated the actual subsurface of the areas covered.

Some GPR operators favor standardizing terms for certain signal patterns. Although this could be considered helpful when comparing one part of a site with another, it may also be misleading since similar patterns may be pro-duced by different aggregates of buried material on different sites.

If ground truthing is carried out before the GPR report is issued, refer-ence should be made to this in the report and, if possible, any relevant evi-dence presented alongside the GPR data. Where this has not happened and either the GPR results are ambiguous or subsurface investigation is both pos-sible and potentially useful, this can be included in the recommendations for further work.

SURVEY FAILURES

Where the survey has failed to produce some or all of the expected results, there should be an explanation, based on practical on-site experience, which reflects fairly both the work done, the manner in which the survey was designed and carried out, and the conditions that resulted in failure. Where possible, evidence from the data should be used to illustrate the problems encountered alongside the explanation of why the survey has not produced the expected data. It is not sufficient to simply state that the GPR did not work without further clarification.

CONCLUSIONS

The conclusions of the report must reflect the survey work done and the rea-sonable interpretation of the results. It is acceptable to speculate on the results provided that a difference between hard facts and possible inferences is made clear. It may also be helpful to add in recommendations for avoiding distur-bance of certain areas due to the density of subsurface targets and to list any recommendations for future work whether intrusive or nondestructive.

PROFESSIONAL STATEMENT OF INDEMNITY

As for all geophysical surveys, it is important to include a short statement on the reliance which can be placed on the interpretation of the results relative to the known limitations of GPR survey so that readers of the report understand what may not have been detected, and why the results may be open to alternative interpretations. This is known as a statement of indemnity. Where this is placed in a report depends primarily on the GPR application. For example, all drawings based on utility detection normally carry such a statement. Archeological guidance is for this to be included following the report conclusions (English Heritage, 2008; Schmidt et al., 2015). Other GPR surveyors may place this information between the Introduction and Methodology or within the Methodology section.

FURTHER INFORMATION

A contact name should be given for discussion of the results and/or questions arising from the report. Geophysical reports are quite often commissioned by one body but used by others, in which case, matters arising from the survey may not all have been cleared before the final release of the report. It is also quite common, in utility detection for example, for another survey to be commissioned at a later date, following on from the work previously carried out, particularly if this included the insertion of additional utilities into the survey area. In these cases, it is important to have a designated person competent to discuss the contents and conclusions of the report.

ACKNOWLEDGMENTS AND REFERENCES

If the report has made reference to published material, these references should be noted in full at the end of the report. Acknowledgments generally include those for assistance with any aspect of the survey from issues such as site access, traffic control, or provision of a towing vehicle to collaboration on completing the survey. This may also include recognizing the contribution to interpreting results from discussions with one or more experts in an associated field.

APPENDICES

Appendices can be used for plans, drawn sections, and photographs. They may also be used for the data evidence of the report if this is not included within the results section.

THIRD-PARTY ARCHIVES

In addition to the GPR surveyor's own archives, it may be necessary for both data and reports to be stored in a third-party archive. It is often useful for those planning GPR surveys to have access to earlier material in the same area, particularly in the fields of engineering and utility detection. There have been attempts to collate utility plans in online resources but these have not always been successful, partly because of a reluctance to share ownership of data which has an associated cost and partly because of concerns that access to this data could be gained by someone with intent to damage the identified assets.

For archeological material, the results of GPR surveys are of interest to the national archeological service of the country as well as any archeologists working either in the relevant type of site or in the immediate area. It is therefore good practice to make sure that the reports and results of archeological GPR surveys are archived in the relevant database. This is usually the database recommended by the national archeological service. In addition, the full survey data set should be archived with a competent designated authority where it can be accessed by other researchers. There is usually a charge for this storage and often also a prescribed format for the data submission.

The archiving process may involve submission of data and/or reports to more than one database. For example, in the United Kingdom, The Archaeology Data Service (ADS) is the recognized data storage service but other entities interested in GPR reports include the National Monuments Records of England, Scotland, and Wales (three separate entities) and the local authority historic environment records. The Historic England (formerly English Heritage) geophysical survey database is no longer separate, having merged with ADS in 2011. In the United States, Digital Antiquity maintains the digital archeological record (tDAR) to which archeological GPR data may be submitted. Details of international guidance on data archiving can be found by Schmidt (2013).

INAPPROPRIATE REMARKS IN PUBLICATION

There have been some unfortunate comments published, not always in the context of formal reporting, which should never appear in connection with GPR surveys. These include references to the returned signals as representing "denser material." GPR signals are generated by changes in electromagnetic properties between one material and another. They do not and cannot represent density. One might guess that the data interpreter thinks that the strong signal amplitude represents building materials although this has not been explicitly stated and no reason for this conclusion has been given. Although this may be an accurate guess of what lies buried, it is not a valid deduction without supporting evidence, for example, from the signal patterning in horizontal time slices. High signal amplitudes may be from buried building material such as old foundations, cellars, or an old chamber connected to a utility network. However, high signal amplitudes might also indicate the presence of an air gap which would, of course, represent a lower density of materials. Then again there is the possibility of moisture content or the presence of metal. The amplitude indicates the magnitude of a change in electromagnetic properties from one type of material to another. It is not an appropriate interpretation of high amplitude signals to interpret these as a measure of density.

Another unfortunate statement which has appeared in recent years is that GPR does not work in a certain location, typically a named city, presumably because the surveyor is aware of other failed surveys in the same general location. This is both unhelpful and inaccurate and a cynical reader could be forgiven for wondering whether the limitation was due to the site, the equipment, or possibly even the user. Of course, it is entirely possible that GPR will not work in this particular location but the reason for this should be expressed in terms of GPR operation, not simply mentioning the location (cf Chapter 6: Attenuation or Loss). The reasons for failure could be due to unsuitable soils and signal attenuation, complicated results which were obscured by the presence of other material which therefore prevented the resolution of the expected targets. Equally it could be that the radar did work (gave valid, verifiable results) but was not used in the right location to find the target or not set to a deep enough probing depth. This, of course, assumes that the mistake arose from the information originally given to the surveyor rather than the surveyor making some elementary mistakes! Or, since the radar can only detect within its own limits what is actually there, the site has been largely destroyed and therefore left in a state where it is no longer possible to distinguish the target from its immediate environment (cf Pitts, 2014). All of these are valid reasons for failure in the terms of the GPR.

Just because the technique did not work on one particular site does not mean that it will not work on any other site in the same city. If we take London as an example, it is built primarily on clay and some of the deeper foundations are wet, at times waterlogged. This does not mean that it is not possible ever to use GPR in London. There have been successful GPR surveys within the city. There have also been some which were unsuccessful, usually due to the amounts of waterlogged clay in the subsurface. The appropriate reason in this case would be to mention the waterlogged clay as the cause, possibly illustrating this with a typical sample of the data, rather than blaming the location. It is probably also worth considering whether it might be possible to carry out the survey in drier conditions at a different time of year, in which case this could form one of the report recommendations.

In cases of doubt regarding the electromagnetic response of the soil, it is worth taking one or more soil samples and having these analyzed using a network analyzer to see if it might be suitable for GPR survey. This would also have the benefit of illustrating the optimum frequency of antenna and indicating whether drier conditions might make the survey viable (see also Chapter 6: Attenuation or Loss).

One other problem frequently encountered in reports is matching radar results to borehole information. This is virtually impossible to do unless the radar data are recorded first and predetermined positions marked for cores to be taken at a point which can be matched to positions in the GPR data. If the cores are taken first, it is no longer possible to survey at these points. Matching different positions to those taken by the radar is potentially problematic since the cover depths may (and often do) vary across the survey area.

If the survey has been properly planned and if the function of the radar is understood by the operator, there should be relatively few survey failures. Such failures that there are should be attributable to a recognized cause such as the presence of water in combination with an ionized material such as salt. The various problems listed above indicate either a lack of planning or a lack of understanding of GPR technology. If this continues, it tends to diminish trust in what is actually a very reliable technology. Part of the purpose of this book is to bring to the attention of both GPR surveyors and their commissioning clients the basics of how GPR works so that expectations of the technology can be realistic, achievable, and readily understood.

REFERENCES

ASCE38-02, 2002. Standard guideline for the collection and depiction of existing subsurface utility data. Available from <www.asce.org>.

English Heritage, 2008. Geophysical Survey in Archaeological Field Evaluation. English Heritage Publishing, Swindon, UK.

PAS 128, 2014. Specification for underground utility detection, verification and location. British Standards Institution. Available from: <www.shop.bsigroup.com/>.

McCann, W.A., 1995. GPR and archaeology in central London. Archaeol. Prospect. 2, 155−166.

Schmidt, A., Linford, P., Linford, N., David, A., Gaffney, C., Sarris, A., et al., 2015. EAC Guidelines for the use of geophysics in archaeology, questions to ask and points to consider, EAC guidelines 2. Available from <http://european-archaeological-council.org/files/eac_guidelines_2_final.pdf>.

Digital Antiquity website <www.digitalantiquity.org>.

Schmidt, A., 2013. Geophysical Data in Archaeology: A Guide to Good Practice. Oxbow, UK.

Pitts, M., 2014. Digging for Richard III. Thames and Hudson, London, UK.

Chapter 15

Regulation

Chapter Outline

GPRs are transmitting devices and their use is therefore regulated in a world where the number of transmitting devices from Bluetooth-enabled appliances to radio, television, and mobile telephones is increasing rapidly. Most of these devices have mass market appeal throughout the world whereas GPR, although vital in terms of public safety, is used by a much smaller, more specialized group. The impetus behind the regulation is the prevention of interference with all other bandwidth users.

Although GPR occupies a small niche in worldwide market terms, it uses a wide range of radio spectrum: all GPRs are ultra-wideband devices. Unlike most other transmitting devices, GPR uses very low power. However it is important for both users of the technology and users of other transmitting devices that there should be no transmission breaches. There is no universal specification for GPRs but there are regulations in most countries for the performance and operation of the equipment and some requirements for the operator. There are two major licensing regimes: the European Telecommunications Standards Institute (ETSI) equipment licensing rules (which also apply in Africa) and the US licensing rules governed by the Federal Communications Commission (FCC) which are applied throughout North America. The two licensing systems are markedly different.

EUROPEAN LICENSING OF EQUIPMENT

Licensing of equipment in Europe is based around the concept of not causing interference to other users of the spectrum. The testing and operating regimes

Ground Penetrating Radar. DOI: http://dx.doi.org/10.1016/B978-0-08-102216-0.00015-1

depend on the measurement of stray radiation while the equipment is in use. In this context, both ground coupling and screening (or shielding) of the antennas are important concepts while the equipment is transmitting. Although it is not essential for anyone picking up and using a radar to understand the complexities of GPR design and licensing, it is vital that any GPR surveyor is aware that their GPR should never be left transmitting when not in use, should always be used ground or wall coupled or within the recommended height of the survey surface for antennas being used at speed on either the road or the rail networks (see also Chapter 3: More Fundamentals of GPR Operation). Failure to comply with these rules breaks European licensing legislation and leaves the operator liable to prosecution, particularly in the event of interfering with other transmissions.

Equally importantly every GPR surveyor (or commissioner of GPR survey work) should make sure that only equipment which has been tested under the current licensing regulations is used. Although testing can be carried out in-house by manufacturers, there is an agreed, prescribed process for this and manufacturers of equipment marketed and used in Europe are obliged to indicate conformity with European regulations by "CE" marking their equipment. Any radar which is not "CE" marked does not necessarily comply with European regulation and should never be used within Europe. If untested, there is a risk that it does not comply which would make its use unlawful and potentially put the user at risk of prosecution, particularly if interference with other equipment results. It makes no difference where the equipment was manufactured. If it is marketed and/or used within Europe, it must comply with European legislation.

The original regulations introduced for the legal operation of GPR were:

- The Radio and Telecommunication Terminal Equipment (R & TTE) Directive (still valid until the end of the current transition period, June 2017); and the harmonized standards which flow from this including
- ETSI standard EN 301 489-32 demonstrating compliance with electromagnetic compatibility requirements; and
- ETSI standards EN 302 066-1 and EN 302 066-2 in respect of conformance with spectrum requirements.

In order to take account of the changing specifications of all transmitting equipment, European standards are revised at regular intervals. A new European directive to replace the R & TTE directive has been agreed by the various national administrations as of December 23, 2016. The same principles in respect of the avoidance of stray radiation which could potentially interfere with other devices are still being applied. The principal change is in the technical specification of the receiver design, specifically its efficiency, and the manner in which this is tested. The new radio equipment directive (ReD) has been in place in its transitional phase from June 2016. For the 1-year transition period, the GPR equipment used can comply with either licensing specification. The revised ETSI standard 302 066 is expected to be published in the Spring 2017 official journal. From June 2017, all GPR equipment produced for use within Europe must meet the new specification, even if the design is an existing product. Previously manufactured equipment is not affected. The responsibility for compliance in testing rests with the manufacturers, not the equipment users. However all users and those who commission them should be aware that there is a change of testing regime in order for properly regulated equipment to be eligible for CE marking. Equipment which is not CE marked should not be used within Europe (or Africa).

Full details of current licensing requirements can be found on the website of the European GPR Association (EuroGPR).

Unscreened antennas will not meet the CE requirements under either the former or the current testing regime. It is, however, possible to use these under certain exceptional circumstances. Manufacturers should apply to their national licensing authorities (the National Regulator) for a development license. Users must similarly apply to the National Regulator in the European country where the equipment is to be used for permission to use these radars. Typically the use will have to be justified, the location given, and the license would normally be temporary. A similar process should be observed by anyone wishing to use radars whose frequency band falls outside the approved range. In recent years National Regulators have not refused reasonable requests to use GPR equipment for legitimate surveying purposes provided that there is no objection from sensitive sites in the region.

OPERATOR REQUIREMENTS

There is no pan-European agreement covering requirements for licensing of GPR operators. The ECC decision of December 1, 2006, on the use of the radio spectrum by Ground and Wall probing radar imaging systems allows each country's National Regulator to set these. Most countries have not put a specific regulatory framework in place but accept the general use of GPR on the assumption that it is being used in a professional manner. A few countries have introduced a licensing system. This is generally free with the exception of the United Kingdom where the regulator Ofcom makes a nominal charge for a 3-year license.

GPR OPERATIONAL REQUIREMENTS: THE CODE OF PRACTICE

The relatively relaxed attitude to operator licenses reflects the lack of significant transmission breaches caused by GPR use to date. However GPR users need to be aware that the overall context of the licensing regulations has resulted in the adoption of a European Code of Practice which includes directions for general operation as well as how to proceed when working in the vicinity of sensitive sites. If a transmission breach were to occur and the radar operator were proved not to have complied with the provisions of the Code of Practice, the potential implications for the user are serious.

The origins of the European Code of Practice lie with the EuroGPR which, in order to minimize the risk of its members' GPRs interfering with other equipment operating in the radio frequency spectrum, introduced a Code of Practice for its members. This Code of Practice has since been adopted and endorsed by ETSI as *ETSI EG 202 730*, electromagnetic compatibility and radio spectrum matters (ERM), Code of Practice in respect of the control, use and application of Ground Probing Radar (GPR), and Wall Probing Radar (WPR) systems and equipment. The ETSI Code of Practice now applies to all users of GPR equipment within Europe. The full text can be found on either the EuroGPR or the ETSI websites. It deals with the proper use of GPR, including on or near sensitive sites.

In terms of general use, operators are required to follow the manufacturer's instructions in the proper use of the equipment and to apply a number of measures which will minimize the chances of stray radiation causing interference. These are:

- Switching on GPR equipment only when it is being used for measurement
- Having and using a GPR deactivation mechanism (which can be either hardware or software)
- Keeping the equipment ground or wall coupled during use, wherever practical
- Checking for sensitive sites in the area of operation in advance of a planned survey.

For wall probing systems, the thickness of the wall should be sufficient to absorb the radiation emitted by the radar. Obviously it is not practical to keep GPRs that are used at speed ground coupled on either road or rail networks but these should be kept as close to the surface as possible. Practical experiment suggests that a depth of the order of one-eighth of a wavelength should give good transmission into the ground, and so minimizing stray radiation (see also Chapter 3: More Fundamentals of GPR Operation).

Typically sensitive sites are defined as defense sites, airports, radio astronomy sites, and prisons although there is also provision for National Administrations to define their own list. Two further categories which it is wise to be aware of are railway companies, usually sensitive about the potential for interference with their signaling systems, and hospitals, especially intensive care units. The Code of Practice recommends that where a survey is to be carried out within 1.5 km of a sensitive site, the surveying company should liaise with that site to obtain agreement for use of its GPR equipment.

Although its status is European Guidance rather than mandatory, it would be advisable for all GPR surveyors to be aware of the contents of this short document and also to be able to prove that its guidance had been complied with in order to prevent the occurrence of a transmission breach occurring.

GPR OPERATIONAL REQUIREMENTS: THE RADAR LOG

The Code of Practice has also adopted the use of a Radar Log, another EuroGPR initiative, introduced in order to demonstrate the low likelihood of a properly used GPR interfering with other transmitting devices. This log should ideally be kept in electronic format and should show the following information:

- A map reference for the site
- A brief description of the measurements performed
- Details of the equipment used (i.e., the manufacturer, the model, the serial number, and the nominal frequency of all antennas used)
- The time and date of operation.

The preference for an electronic format is to make the process of gathering in information and comparison with the details of a transmission breach easier for National Regulators. EuroGPR members submit their logs through the Association.

EUROPEAN TELECOMMUNICATIONS STANDARDS INSTITUTE

As its name suggests ETSI's role in licensing is concerned with prescribing the standards both for testing and for operation of telecommunications equipment, including GPR. In addition to its regulatory role, ETSI supports the industries it recognizes by producing information booklets on relevant industries. The information booklet on GPR/WPR is available either directly from ETSI or from EuroGPR (Fig. 15.1).

THE EUROGPR AND ITS ROLE IN LICENSING MATTERS

Unlike American provisions, European licensing regulations contain provision for review and revision (if required) at regular intervals. One of the easiest ways of staying up to date with developments in licensing whether of equipment or of operators is membership of the EuroGPR. Under the leadership of the licensing officer, EuroGPR takes an active part in revisions of the European licensing framework and continues to cooperate with the various European bodies who collaborate over licensing matters. EuroGPR has been a member of ETSI since March 2007. Although, as the name suggests, the association is based in Europe, all major manufacturers including those based in North America, are members and actively participate in all technical licensing specifications.

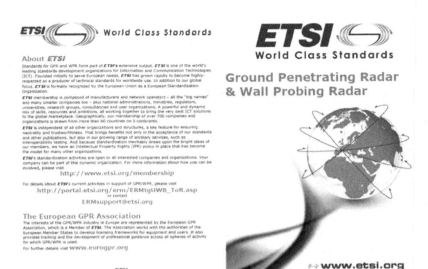

FIGURE 15.1   ETSI booklet promoting GPR and WPR.

EuroGPR grew out of the UK-based Impulse Radar Users Association whose original aims were to create an appropriate context for the use of GPR, taking into account the wide bandwidth, the low power, and the lack of interference with other spectrum users. As a UK-based association, their original discussions were primarily with the UK regulator of the time, the Radiocommunications Agency (RA). However, in 1994, GPR was used in a very public arena, to detect the remains of a number of young girls who had been tortured, murdered, and buried by Fred West, a Gloucestershire-based builder and his wife, Rose. Since there was, at the time, no legal context for GPR as transmitting equipment, it quickly became apparent that there was a potential conflict between the RA, as the official regulator, and the Home Office who, through the medium of the police force, wished to use GPR technology and refused to have their investigations impeded by an apparent lack of legal context for the GPR equipment. The United Kingdom's interaction with Europe on the political level also made it clear that the UK-based organization needed to expand and join forces with their mainland European colleagues. The pan-European need for the technology was underlined a few years later with the GPR search in Belgium for the remains of Marc Dutroux's victims (Daniels, 2004; Groen et al., 2015).

In July 1997, European GPR Association was formed with early membership coming from companies based in Sweden, the Netherlands, and Germany. The Association now covers 14 European countries and also has members in the United States, Canada, and Nigeria. One central aim was to ensure that the relatively niche market of GPR was not squeezed out of their use of the spectrum by other technologies which, although not competitors, make use of parts of the same spectrum (e.g., mobile phones, GPS, the satellite industry). When the first FCC regulation in the United States came into being in 2002 with some perceived disadvantages to the GPR industry, EuroGPR was encouraged to become as involved as possible in the decisions being made at European level. Initial representation was by two members, one European manufacturer and another manufacturer's representative. In 2004 increasing membership resulted in a reorganization and on March 28, 2007, EuroGPR became a full member of ETSI, giving the association direct access into the technical discussions which form the GPR licensing context in Europe.

In addition to the Code of Practice referred to above, all Association members sign up to a Code of Ethics. The Association also provides GPR Application Guidelines which are available to Members via its website. From its sponsorship of international conferences, EuroGPR has built a large virtual library of conference papers, recommended reading by topic, and recommended reading by author. It is in the process of developing an Internet-based training program for users of the technology. Membership of the organization is not mandatory but can bring benefits in terms of representation at European level, guidance on technical issues, and the opportunity to interact with fellow professionals. It is also a means of demonstrating the intent to act professionally and responsibly (cf Appendix B: Professional Association and Reputable Manufacturers). The Association aims to develop branch organizations within other European countries and has recently formed an Italian section.

NORTH AMERICAN EQUIPMENT LICENSING

The American licensing system was developed by the FCC together with the National Telecommunications and Information Administration (NTIA) in 2002. The FCC rules were introduced from many of the same concerns as in Europe, i.e., the potential for conflict in the use of parts of the radio frequency spectrum between GPR users and others. It differs from the European system in that there are limitations placed on the levels of radiated signal which vary with frequency. For certain frequencies which are deemed critical to other users (e.g., GPS) these are restrictive, especially relative to the European permitted designs.

From the introduction of the regulations, all GPR equipment has been required to be registered with the FCC. This applied initially to all systems owned prior to the introduction of the rules, with a deadline for registration of October 15, 2002. All GPRs manufactured or introduced into the United States since that date require to be tested by a laboratory, independent of the manufacturer, to ensure compliance with the regulated emission levels. Then the manufacturer submits the test results and a detailed description of the system to the FCC for certification. Assuming that this is successful, the FCC issue an ID number which is unique to that system (or part of the system, e.g., a new antenna).

All equipment owners who purchased equipment from the July 15, 2002, up to and including the present day are required by federal law to register their certified radar equipment with the Frequency Co-ordination branch of the FCC using the relevant ID number for the system(s). Forms for this purpose are available from GPR manufacturers as well as the FCC. In addition to the certification number, the information requested includes ownership and contact details, areas of operation, make and model of the GPR, and the center frequency of all antennas purchased.

Detailed information on the licensing system can be obtained from the FCC and NTIA websites. From the point of view of the user, the critical element is to make sure that any equipment you are using has been registered with the FCC and carries the FCC identification number.

In 2009, after wide consultation which included inviting comments from the FCC and from EuroGPR, as well as their own radio spectrum users, the Canadian National Regulator, Industry Canada, issued their mandatory document RSS 220 "Devices Using Ultra-Wideband (UWB) Technology." This regulation follows the FCC regulation rather than the ETSI one, making the regulatory position coherent across North America.

GPR AND SAFETY OF TRANSMISSIONS

The potential safety issues associated with radiation came to public notice in connection with concerns over the use of mobile phones although laptop computers have similar strength fields of radiation to the latter. There are few safer technologies than GPR/WPR. This is primarily because of the very low power used by these devices. Shielding of the antennas also protects the user as well as protecting the GPR data from spurious external aboveground signals and using the equipment ground coupled, as it should be, is a further protection.

According to the International Commission on Non-Ionising Radiation Protection (ICNIRP), the safe power flux density limits for the general public are 2 W/m^2 rising to 10 W/m^2 for a frequency range of $<400 \text{ MHz}$ up to 2 GHz and above. At these frequencies, the limitation is the heating of human tissue. The limits for occupational users are five times higher. The radiation from GPR and WPR lies well below these limits and the safety risk from using GPR/WPR is therefore negligible as can be seen from the following example where the calculated flux density limits are compared with the safe public limits.

If, instead of being used in an appropriate manner, the GPR were used for scanning directly on to a human body, the following flux densities would be the result. For typical GPRs with impulse transmitters, at 10 MHz, 400 MHz, and 1 GHz, the resulting levels are:

- At 10 MHz: 9 mW/m^2 which would be 200 times below the public limit
- At 400 MHz: 4 mW/m^2 which would be 500 times below the public limit
- At 1 GHz: 2 mW/m^2 which would be 2500 times below the public limit (Utsi, 2015).

Although this demonstrates that the radiation risks from GPR/WPR are negligible, this table should not be taken as an indication that it is acceptable to radiate a GPR directly on to a person. It should be emphasized that GPR should never be applied directly to a human body, except for the limited cases where it is being used for scanning and detection of existing tumors. It is simply unprofessional not to use the equipment for the purpose intended and in the prescribed manner. It also cuts against both professional standards and the legal requirements of the regulation which in Europe, for example, aims not to breach any other transmissions (cf also Persico et al., 2015).

However, it is worth noting that the power radiated by a GPR falls well below the existing safety limits, even if the radar were to be used in this unapproved, inappropriate fashion. The above results are, of course, why it is possible to use GPR for detection directly on to the human body, under certain circumstances. Since it demonstrates the theoretical position without the normal benefits of either screening or ground coupling to a normal survey surface, it underlines the inherent safety of the technique when used in a professional and legitimate manner.

OTHER ORGANIZATIONS

There is increasing interest by other professional bodies in ensuring that GPR work is carried out to a professional standard and that work quoted for will conform to a widely recognized specification. The American Society of Civil Engineers (ASCE) were one of the earliest organizations to make a formal specification of the requirements of different levels of utility detection surveys, including the role of GPR. In the United Kingdom, the Survey Association (TSA) put together guidance on utility surveying which included some descriptions of mainstream GPR equipment as well as expected input for various levels of utility detection. They are currently exploring the possibility of introducing an accreditation system to be run by Lloyds register, to include an assessment of competence to carry out GPR surveys.

The UK's Geological Society also has many members with professional expertise in GPR and will consider GPR-based projects as part of an application for qualification for chartership.

In Japan an organization similar to EuroGPR has recently been formed. Although the association is still relatively small, a committee has been formed to liaise with government over the allocation of GPR frequency bands.

Within North America, there is no equivalent to EuroGPR but professional users of GPR equipment are frequently members of professional organizations such as the Society of Exploration Geophysicists, the American Geophysical Union, and the Environmental and Engineering Geophysical Society, all of whom include professional GPR use in their interests and publications.

REFERENCES

ASCE38-02, 2002. Standard guideline for the collection and depiction of existing subsurface utility data. Available from <www.asce.org>.

Canadian RSS 220. Available from <https://www.ic.gc.ca/eic/site/smt-gst.nsf/eng/sf09347.html>.

Daniels, D.J., 2004. Ground Penetrating Radar, second ed. The Institution of Electrical Engineers, London, UK, pp. 4, 429–433.

European GPR Association (EuroGPR). <www.eurogpr.org>.

European Telecommunications Standards Institute (ETSI). <www.etsi.org>.

Federal Communications Commission. <www.fcc.gov>.

Groen, W.J.M., Márquez-Grant, N., Janaway, R.C. (Eds.), 2015. Forensic Archaeology: A Global Perspective. Wiley, Oxford, UK, pp. 11, 192–193.

NTIA. <www.ntia.doc.gov>.

Ofcom, UK National Regulator. <http://www.licensing.ofcom.org>.

OfW349. GPR licence application form is available from: <http://licensing.ofcom.org.uk/binaries/spectrum/low-short-devices/ground-probing-radar/ofw349.pdf>.

Persico, R., et al., 2015. Recommendations for the safety of people and instruments in ground penetrating radar and near surface geophysical prospecting, EAGE Publications BV3990 DB Houten, The Netherlands. ISBN 9789462821620.

Utsi, V., 2015. Calculation of theoretical flux density of GPR radiation completed for the TU Cost Action 1208.

Glossary of Common Technical Terms

Amplitude See Signal amplitude.

A-scan A single signal trace recorded by the radar receiver. Two-dimensional data is made up of all such signals. The primary value of an A-scan or wiggle trace is monitoring quality control of the data but it is also useful for determining Tzero (see below).

Antenna The part of the radar which is used for transmission or reception of radio waves.

Attenuation Loss of the signals emitted by the radar is caused by the electromagnetic properties of the material through which the radio waves pass, notably its ability to hold charge (relative permittivity), conductivity, and magnetic response. Most soils attenuate the signals to varying degrees. See also Chapter 6, Attenuation or Loss.

Background removal This is a process applied to the data to remove the constant background from the data in order to highlight the location of target material. There are a large number of methods of doing this, ranging from a simple averaging process over the full data set to selection of particular areas as representing the common environment. See Chapter 8, Data Processing.

B-scan This is another term for two-dimensional data, also sometimes referred to as a radar profile, trace, or radargram. The B refers to the fact that each B-scan is made up of a series of A-scans.

Bistatic GPR A GPR using at least one transmitter antenna and a separate receiver antenna although these may be contained within the same antenna box. See also Monostatic GPR.

C-scan The C-scan, also known as a time slice, is a horizontal view extracted from the three-dimensional data which is made up from the collection of two-dimensional data or B-scans.

Crossed dipole antennas Antennas are normally placed in combination so that their e-Fields are aligned. Crossed dipole antennas are usually used for the detection of linear features and they are placed in combination so that their e-Fields are orthogonal to one another as this optimizes the detection of linear features over that of layers: see Chapter 1, Introduction to GPR.

Curve fitting A commonly used method of calibrating transmission velocity by matching the shape of hyperbolas in the GPR data to a theoretical curve based on depth and transmission velocity. See Chapter 5, Velocity Calibration.

Dielectric constant The dielectric constant is a measure of a material's ability to hold electric charge. It is an important factor in the transmission of radio waves. See Chapter 1, Introduction to GPR.

e-Field   The electric field associated with each GPR antenna. This has both value and orientation.

Encoder wheel   A means of measuring the distance travelled by the radar during surveying. This may be an optical device, a magnetic device, or a bridge to another measurement system. Sometimes referred to as an odometer.

Er   See Dielectric constant and Relative permittivity, also Chapter 1, Introduction to GPR.

Frequency   Frequency is a measure of the number of radio waves passing a set point in 1 second and, for GPR, is usually measured in MHz and GHz. 1 MHz is equivalent to 1,000,000 waves per second and 1 GHz is equivalent to 1000 MHz (or 1 billion waves per second).

Gray scale   An option for data presentation where the data is shown in shades of white, black, and gray according to the signal amplitude. There are several options for gray scale presentation.

Ground coupling   Coupling refers to the efficient transmission of signals from the radar into the survey medium (e.g., the ground). It is usually achieved by antennas being balanced electrically (done as part of the electronic design by the manufacturer) to the ground and by the surveyor maintaining as good contact between the antenna and the ground (or other survey medium) as possible. It is also possible to have air-coupled antennas and water-coupled antennas. See also Chapter 3, More Fundamentals of GPR Operation.

Ground truthing   Verification of the GPR results in depth measurement for comparison with the time measurement of the GPR in order to calibrate transmission velocity. The process may involve extraction of cores or limited excavation, depending on the site.

Hyperbola   The basic shape formed by signals returned by GPR targets is a smooth symmetrical curve in one plane known as a hyperbola. See Chapter 3, More Fundamentals of GPR Operation, for the reasons why this is not always the case in practice.

Interference   Unwanted transmissions which appear on the GPR data, sometimes the result of other transmitting devices such as mobile phones, sometimes the result of secondary reflections from (for example) overhead features such as metal beams and power lines. Interference is a potential problem since it may obscure real subsurface data.

Lossy   A soil or other survey medium is said to be lossy when it absorbs a significant proportion of the transmitted signal, reducing the signal available for return to the receive antenna.

Mark up on site   A survey strategy sometimes applied when using GPR for utility detection. The position of detected utilities is marked on the ground during the survey and the GPR data is not normally processed after survey.

Meandering   A method of collecting GPR data by completing parallel survey lines beginning and ending at alternate ends of the survey area. This is usually done to save time on site where the survey area is large enough that walking back to the original start line would be time consuming. It will be necessary to align the radar profiles into the same orientation, particularly if a three-dimensional data block is to be constructed.

Migration   A mathematical technique for reducing the shape of a hyperbola to a discrete point. Migration can be used to define more accurately the position of linear features such as utilities. It can also be used to refine the calibration of transmission velocity. See Chapter 5, Velocity Calibration, and Chapter 8, Data Processing.

Monostatic GPR   A single antenna is used for both transmission and reception. See also Bistatic GPR.

Multichannel GPR   Any GPR which has more than one transmitter and one receiver in operation at the same time. There are a wide variety of multichannels, some of which are also multifrequency.

Nanoseconds   GPRs measure probing depths in nanoseconds time. One second equals 1,000,000,000 nanoseconds (ns).

Noise   Unwanted signals such as ringing (echo effects), spurious effects generated from the survey, or equipment which may disguise or impair the visibility of subsurface signals. See also Chapter 3, More Fundamentals of GPR Operation.

Odometer   A means of measuring the distance travelled by the radar during surveying. This may be an optical device, a magnetic device, or a bridge to another measurement system.

Q   Impedance ratio, a means of evaluating the electromagnetic response of a soil sample using a network analyzer. See Chapter 6, Attenuation or Loss.

Radar profile   A radar profile, trace, or radargram is a line of GPR data depicting the depth of signals below the path travelled by the radar. It is also known as two-dimensional data.

Radar trace   A radar profile, trace, or radargram is a line of GPR data depicting the depth of signals below the path travelled by the radar. It is also known as two-dimensional data.

Radargram   A radar profile, trace, or radargram is a line of GPR data depicting the depth of signals below the path travelled by the radar. It is also known as two-dimensional data.

Relative permittivity   Relative permittivity is a measure of any given material's ability to hold charge. This is an important factor in the transmission of radio waves. See also Chapter 1, Introduction to GPR.

Ringing   Ringing is the technical term for echo effects. These may arise from certain targets such as metal artifacts or voids. However, all radars also have a certain amount of background ringing in their design, normally not visible due to the returned signals. Visibility of the background ringing is an indicator that signals are being lost into the environment.

Sender/receiver correction   Where the transmitter and receiver antennas are separate entities, there is a physical gap between them which means that the two-way travel time is not equal to twice the depth of the target, especially for shallow targets. The S/R correction is to correct the geometry. See Chapter 8, Data Processing.

Signal amplitude   The amplitude is the strength of the signal, measured as the amount of disturbance of the wave from a neutral position. Note that this does not mean from full negative to full positive but from the zero position to either the maximum positive or maximum negative reading. Fig. 3.1 illustrates this.

Signal-to-noise ratio   Signal is the wanted part of the information in a GPR radargram, whereas noise is effectively interference. A high signal-to-noise ratio is the aim of most GPR design. See also Chapter 3, More Fundamentals of GPR Operation.

Single channel GPR   A GPR which may make use of different antennas but has a maximum capacity of one transmitter and one receiver at any given time. Alternative frequencies or types of antenna would have to be used sequentially, not simultaneously.

Stacking   A method of averaging over a number of traces in order to improve the signal-to-noise ratio of the data.

Swept frequency Most GPRs maintain a fixed frequency range and sweep or probe through time (depth). Swept frequency GPRs maintain fixed time but sweep through a range of frequencies.

Three-dimensional data The two-dimensional data from each survey line can be combined into a three-dimensional data block from which, typically, horizontal views may be extracted in order to examine the signal patterning.

Time slice A horizontal view extracted at a common time from a three-dimensional data set built up from a series of radargrams. Where there are mixed materials especially if these include either air or water so that the transmission velocity varies, this may be a quasi-horizontal view unless correction is made for the variation in velocity.

Time zero (Tz) The time in nanoseconds for the transition of the signals from the generation point into the ground via the transmit antenna. As this does not represent the subsurface, Tz requires to be removed from the data in order for the target depths to be accurately measured.

Transducer A device which translates one form of energy into another. In GPR technology, this can refer to either a transmit antenna or a receive antenna. A transducer pair normally refers to a transmitter plus a receiver. See Chapter 1, Introduction to GPR.

Transect spacing The physical distance between successive survey lines.

Transmission velocity Radio waves do not travel at a constant velocity. The speed at which they pass through a given material depends upon the electromagnetic response of that material.

Two-dimensional data The data generated initially by GPR shows the depths of signals plotted against the distance travelled by the radar. This is the normal two-dimensional data, also referred to sometimes as "B-Scan," radargram, radar profile, or trace.

Two-way travel time The radar's measurement of depth is from the transmit antenna to the target plus from the target back to the receive antenna. This measurement is approximately twice the depth of the target except at shallow depths. See also Chapter 8, Data Processing.

WARR Wide angle reflection and refraction, a method of calibrating transmission velocity by separating the transmit and receive antennas during transmission. Since the transmission velocity through air is known (0.3 m/ns), the transmission velocity through the ground can be derived. See also Chapter 5, Velocity Calibration.

Wavelength The distance between one point on the waveform and the identical point on the following waveform. Fig. 3.1 illustrates this.

Wiggle trace A single signal trace recorded by the radar receiver. Two-dimensional data is made up of all such signals. The primary value of an A-scan or wiggle trace is monitoring quality control of the data but it is also useful for determining Tzero (see above).

Appendix A

Further Reading

GROUND PENETRATING RADAR - GENERAL

This book is intended as a beginner's guide and therefore does not go into the science behind GPR in any detail. The essential text book for GPR which does cover this ground is

Daniels, D.J. (Ed.), 2004. Ground Penetrating Radar—2nd Edition. Institution of Electrical Engineers, London, UK. ISBN 0 86341 360 9.

This also gives additional reading references and there are chapters on different applications of GPR. Both of these features are extremely useful background reading even if the reader is not interested in pursuing the finer points of GPR science.

Other useful references are as follows:

Annan, A.P., 2004. Ground Penetrating Radar Principles, Procedures and Applications, Ontario. Also the Chapter by the same author on GPR in Butler, D.K. (Ed.), 2005. Near Surface Geophysics (Investigations in Geophysics No. 13). Society of Exploration Geophysicists, Tulsa, OK, USA.

Jol, H., 2009. Ground Penetrating Radar: Theory and Applications. Elsevier, London, which is also available as an E-book.

A recent additional aid to survey planning is available at http://www.gpr-parameters.ch/about.html (Leckebusch, 2016). This website deals with all aspects of GPR parameters and is extremely useful.

GPR ENGINEERING

There are usually many papers on engineering matters in the proceedings of the international conferences both the GPR and IWAGPR series (see Conference Papers below). The International Structural Faults and Repair conference also has a nondestructive testing module which usually features GPR application papers (see Conference Papers below). In addition, the specialist journals also sometimes cover case studies (see Journals below). Examples include:

Hugenschmidt, J., Herlyn, A., 2014. Damage in pavements caused by previous excavation work? In: Proceedings of the 15th International Conference on Ground Penetrating Radar, Brussels, Belgium, pp. 882–886.

Hamrouche, R., Saarenketo, T., 2014. Improvement of a coreless method to calculate the average dielectric value of the whole asphalt layer of a road pavement. In: Proceedings of the 15th International Conference on Ground Penetrating Radar, Brussels, Belgium, pp. 929–932.

An early practical guide, written in accessible English, was published by the UK Building Research Establishment in 1998, funded by the European Union. The reference below is Volume 1 of the set.

Matthews, S.L., 1998. Application of Subsurface Radar as an Investigative Technique. BRE Garston, Watford, UK. ISBN 1 86081 2104.

GPR UTILITY DETECTION

It is unusual for papers or books to be published on this subject which is rather ironic, given that this is the single biggest use of GPR worldwide. Chapter 13 of Daniels (see General above) deals with utility detection including some case studies from experienced practitioners.

That said, there is a growing body of standards being developed for utility detection. This covers all aspects and not just GPR but anyone intending to use GPR for this purpose needs to be aware of the content. ASCE38-02 is the appropriate reference for the United States and PAS128 for the United Kingdom.

ASCE38-02, 2002. Standard Guideline for the Collection and Depiction of Existing Subsurface Utility Data. Available at: www.asce.org.

PAS 128, 2014. Specification for Underground Utility Detection, Verification and Location. British Standards Institution. Available at: www. shop.bsigroup.com/.

HSG47, 2014. Avoiding danger from underground services. Publ UK Health and Safety Executive, 3rd edition. ISBN 978 0 7176 6584 6.

The Survey Association: http://www.tsa-uk.org.uk/updated-essential-guide-to-utility-surveys/.

"Third Party Damage to Underground and Submarine Cables" at www. landsnet.is/uploads/1067.pdf.

There is a published study of the potential for in-sewer GPR published by the UK Water Industry Research group (UKWIR).

Faragher, E., 2007. Minimising street works disruption. UKWIR Report No. 07/WM/12/16, UK.

There is also a short case study of utility detection using GPR, including the detection of telecoms cable below ground which had been made up to a height of over 6 m in places in the 2004 GPR Conference papers.

Utsi, E., 2004. The use of ground penetrating radar as a risk mitigation tool. In: Slob, E., Yarovoy, A., and Rhebergen, J. (Eds.), Proceedings of the Tenth International Conference on Ground Penetrating Radar, Vol. II, pp. 795—798. ISBN 90 9017959 3.

ENVIRONMENTAL GPR INVESTIGATIONS

Bristow, C.S., Jol, H.M. (Eds.), 2003. Ground penetrating radar in sediments. Geological Society London Special Publication 211. ISBN 1 86239 131 9.

This book includes advice on data collection and processing from two very experienced GPR practitioners. The remainder of the book has papers on work done in a range of environments, e.g., Aeolian and coastal, fluvial and alluvial, glacial, ancient sediments and tectonics. It is clearly written, well-illustrated but, as would be expected, contains technical environmental terms.

A good practical example of using GPR for an environmental study where intrusive methods of investigation could not be used is:

Bakker, M.A.J., 2005. The internal structure of Pleistocene push moraines. A multi-disciplinary approach with emphasis on ground-penetrating radar. TNO Built Environment and Geosciences Geological Survey of the Netherlands. Alphen aan den Rijn, Netherlands. ISBN 90 9020270 6.

For those interested in peatlands:

Utsi, E., 2001. The investigation of a peat moss. In: Vertrella, S., Bucci, O., Elachi, C., Lin, C., Rouzé, M., Sato, M. (Eds.), Remote Sensing by Low Frequency Radars Workshop 2001. European Association of Remote Sensing Laboratories, Naples.

Environmental investigations usually feature heavily in the GPR and IWAGPR Conferences (see below).

For example:

Bunting, C., Branch, N., Robinson, S., Johnes, P., 2014. Ground penetrating radar as a tool to improve heritage management of wetlands. In: Proceedings of the 15th International Conference on GPR, Brussels, Belgium.

Bristow, C.S., 2004. GPR in sediments: recent advances in stratigraphic applications. In: Slob, E., Yarovoy, A., Rhebergen, J. (Eds.), Proceedings of the 10th International Conference on Ground Penetrating Radar, Delft, Netherlands, pp. 569—572.

Markus, T., Green, J.L., Cooper, J.F., Fung, S.F., Taylor, W.W.L., Benson, R.F., Gogeneni, S.P., Ramasami, V.C., Reinisch, B.W., Song, P., 2004. Ground penetrating radar simulations for Jupiter's Icy Moons. In: Slob, E., Yarovoy, A., Rhebergen, J. (Eds.), Proceedings of the 10th International Conference on Ground Penetrating Radar, Delft, Netherlands, pp. 789–792.

GPR FORENSICS

Cheetham, P., 2005. Forensic geophysical survey. In: Hunter, Cox (Eds.), Forensic Archaeology: Advances in Theory and Practice. Routledge, London, pp. 62–95. ISBN 0 415 27311 0; 0 415 27312 9.

This book is aimed at Forensic Archeologists but contains one chapter on a range of geophysical methods, including a short section on GPR. Although, as with much archeological guidance, the focus tends to be on the possibility of detecting a type of archeological feature, in this case graves, rather than looking for the components within the feature which make GPR detection a possibility.

Groen, W.J.M., Márquez-Grant, Janaway, R.C. (Eds.), 2015. Forensic Archaeology: A Global Perspective. Wiley Blackwell, UK. ISBN 9781118745984.

This is also primarily aimed at Forensic Archeologists but has a useful list of references to where GPR has been used, country by country, for forensic purposes and also includes some data examples.

There is a biennial conference on forensics held in London in the Geological Society by the European Association of Geoscientists and Engineers (EAGE), on the day following the archeological conference hosted by English Heritage at the same venue. The date is usually in December on even numbered years.

GPR ARCHEOLOGY

Anyone interested in archeological geophysics is strongly recommended to look at the papers from the International Conference on Archaeological Prospection. The latest of these, the 11th, was held in Warsaw in 2015 and for which the proceedings are available via www.iaepan.vot.pl/ap2015/. The conferences are organized by the International Society for Archaeological Prospection (ISAP)—see also Websites below.

Good general references include the following:

Goodman, D., Piro, S., 2013. GPR Remote Sensing in Archaeology. Springer-Verlag, New York.

Conyers, L.B., Goodman, D., 1997. Ground-Penetrating Radar: an Introduction for Archaeologists. Altamira Press, Walnut Creek, CA, USA.

Conyers, L.B., 2004. Ground-Penetrating Radar for Archaeology. Altamira Press, Walnut Creek, CA, USA. ISBN 0-7591-0773-4.

Conyers, L.B., 2012. Interpreting GPR for Archaeology. Altamira Press, Walnut Creek, CA, USA.

Schmidt, A., 2013. Geophysical Data in Archaeology, Oxbow.

There is a biennial conference on archeological geophysics hosted by English Heritage at the Geological Society in London. The date is usually in December on even numbered years. Papers from this conference usually appear in Archaeological Prospection (see Journals below).

GPR DATA PROCESSING

There are always a good number of processing papers presented at the GPR and IWAGPR conferences (see Conference Papers below). It is advisable for beginners to seek training in any software package they intend to use.

A recent publication dealing with GPR processing is

Persico, R., 2014. Introduction to Ground Penetrating Radar. Inverse Scattering and Data Processing. Institute of Electronics and Electrical Engineers Press, John Wiley & Sons, Hoboken, NJ. ISBN 978-1-118-30500-3.

GPR IMAGING

This is often, but not always, associated with searches for Unexploded Ordnance. This is also a major topic for both the GPR and IWAGPR conferences (see Conferences below). Although many of the papers are relatively advanced, they also include case studies or experiments with obvious practical application. A good example of what can be achieved in this field is

Sato, M., Takahashi, K., 2014. Optimization of data sampling and image reconstruction by GPR. In: Proceedings of 15th International GPR Conference, Brussels, pp. 634–637.

UNUSUAL GPR TECHNICAL CLAIMS

Literature dealing with technical claims which could reasonably be tested by a blind technical test includes the following papers:

Daniels, D.J., Utsi, E., 2013. GPR case history and known physical principles. In: Proceedings of the 7th International Workshop on Advanced

Ground Penetrating Radar (IWAGPR 2013), Nantes, France, July 2013. ISBN 978-1-4799-0937-7; IEEE Catalogue No. CFP13538-PRT.

Hodges, G., 2005. Voodoo methods: dealing with the dark side of Geophysics. In: Proceedings of the Symposium on the Application of Geophysics to Engineering and Environmental Problems (SAGEEP), Atlanta, USA, April 2005.

Hodges, G., 2011. Commentary: There's a dark side to Geophysics. Northern Miner, Canada, 6th May 2011.

McCann, W.A., 1995. GPR and Archaeology in Central London. Archaeological Prospection, Vol. 2, pp. 155−166, 1995.

Olhoeft, G.R. Quality control in Geophysics. In: Geophysical Technologies for Detecting Underground Coalmine Voids, 28−30 July 2003, Lexington, KY, proceedings CD-ROM, 7 p.

Tuley, M.T., Ralston, J.M., Rotondo, F.S., Andrews, A.M., Rosen, E.M., 2002. Evaluation of EarthRadar unexploded ordnance testing at Fort A.P. Hill, Virginia. Institute for Defense Analyses, Alexandria, VA, Aerospace and Electronics Magazine, IEEE, Vol. 17, Issue 5, pp. 10−12.

JOURNAL PAPERS

In general, GPR papers are published alongside other geophysical investigation methods which form part of a geophysicist's toolkit.

Geophysics is an international journal, produced by the Society of Exploration Geophysicists which appears every 2 months. See library. seg.org/journal/. It is peer reviewed and deals with all branches of geophysical exploration, not just GPR, including research and case study materials.

Journal of Applied Geophysics (journals.elsevier.com/journal-of-applied-geophysics) had its origin in the mining industry and now covers developments and innovative applications in environmental, engineering, and hydrological investigations.

Near Surface Geophysics is an international journal published six times a year by the European Association of Geoscientists and Engineers (EAGE). The content is research and development in geophysics in the near surface and therefore covers most applications of GPR. Although expected to be scientifically rigorous, the language is intended to be clear to nonspecialists. GPR articles are frequently included.

First Break, also published by EAGE, appears monthly. Its primary focus is on applications relevant to the oil and gas exploration but it does sometimes cover other geophysical investigation and also carries regular news features.

Archaeological Prospection is an international peer-reviewed journal published four times a year by John Wiley & Sons, UK. As the name suggests the papers published are concerned with Archeological Geophysics and the journal is associated with ISAP (see below).

GPR CONFERENCE PAPERS

There is an international GPR conference every year. On even numbered years, this is the GPR conference, the most recent one being GPR 2016, **the 16th International GPR conference**, held in Hong Kong in June 2016. The 17th Conference will be held in June 18–21, 2018, in Rapperswil, Switzerland.

The previous conferences in this series were as follows.

1st	1986	Tifton, GA, USA	9th	2002	Santa Barbara, CA, USA
2nd	1988	Gainesville, FL, USA	10th	2004	Delft, Netherlands
3rd	1990	Lakewood, CO, USA	11th	2006	Columbus, OH, USA
4th	1992	Rovaniemi, Finland	12th	2008	Birmingham, UK
5th	1994	Kitchener, ON, Canada	13th	2010	Lecce, Italy
6th	1996	Sendai, Japan	14th	2012	Shanghai, China
7th	1998	Lawrence, KS, USA	15th	2014	Brussels, Belgium
8th	2000	Gold Coast, Australia	16th	2016	Hong Kong, China

Many of the papers from these conferences are available online from the American Institute of Electrical and Electronic Engineers (IEEE). EuroGPR hosts a number of conference papers, notably from conferences sponsored by the Association.

On the odd numbered years in between are the **International Workshops on Advanced GPR (IWAGPR)**. IWAGPR 2017, the 9th IWAGPR will be held in Edinburgh.

1st	2001	Delft, Netherlands	2nd	2003	Delft, Netherlands
3rd	2005	Delft, Netherlands	4th	2007	Naples, Italy
5th	2009	Granada, Spain	6th	2011	Aachen, Germany
7th	2013	Nantes, France	8th	2015	Florence, Italy
9th	2017	Edinburgh, UK			

Both conferences are usually hosted by a university. The 2008 conference held in Birmingham, UK, was also co-hosted by EuroGPR. EuroGPR frequently sponsors these conferences and, in return, usually receives the papers into its virtual library (see Websites below).

International Conference on Structural Faults and Repair. This primarily engineering conference is usually hosted in Edinburgh with occasional forays into London, UK. Papers from the 16 International Conferences on Structural Faults and Repair contain some interesting GPR papers on applications and research.

International Conference on Archaeological Prospection includes papers on the use of GPR for archeological investigations. The 11th international conference was held in Warsaw in September 2015.

WEBSITES

Manufacturers. Most of the major reputable manufacturers are members of EuroGPR and can be identified by the letter "M" on the EuroGPR website (see below) and many, if not all, also have direct links to their websites. There is an increasing amount of helpful technical advice on most of these.

COST action 1208 (www.cost.eu/COST_Actions/tud/TU1208). This is a European funded international co-operation project on Civil Engineering Applications of Ground Penetrating Radar. There are publications available by download from the site on civil engineering applications, the progress of research work, and the safety implications of using GPR.

EuroGPR (www.eurogpr.org). This website not only has guidance on the current European licensing position but is also developing a series of Guidelines for specific GPR applications. See "Rules and Regulations."

Members can access, through the virtual library, a Recommended Reading List, papers from GPR conferences and two anthologies of helpful references by subject matter and by author. This includes a list of the papers that experienced GPR scientists consider to be their most important and also recommendations, by subject matter, of good source material. The virtual library may be accessed via "Scientific Literature."

The Association is also developing an online training course.

International Society for Archaeological Prospection (ISAP) (www.archprospection.org). The association publishes a quarterly newsletter for archeological geophysicists, hosts the Europae Archaeologiae Consilium Guidelines for the use of Geophysics in Archeology, and offers reduced subscriptions for Archaeological Prospection (see above). There is also a reasonably active interchange of views and advice via e-mail.

The Chartered Institute for Archaeologists (CIfA) (www.archaeologists.net). CIfA publishes guidelines on all aspects of Archeology, including geophysics. They also run a special interest group for geophysics (GeoSIG).

The Survey Association (TSA) (www.tsa-uk.org.uk) is a trade association representing a number of surveying professions including surveying of the subsurface.

SCIENTIFIC PUBLISHERS

www.elsevier.com
www.interscience.wiley.com

Appendix B

Professional Association and Reputable Manufacturers

FINDING A REPUTABLE MANUFACTURER

Most reputable major manufacturers, wherever they are based in the world, are members of the European GPR Association (EuroGPR) (see also Chapter 15: Regulation). This is partly because the GPR market is a niche market and therefore almost all manufacturers operate on a worldwide basis. The principal North American manufacturers were therefore as concerned as European manufacturers with establishing a suitable and relevant licensing system within Europe and became members who actively participate in licensing issues. All Association members also sign up to a Code of Ethics and a Code of Practice (see Chapter 15: Regulation) which shows, if nothing else, some care for their reputations in terms of technical reliability.

Many of the manufacturers' agents are also members of the Association and their names can be found along with other members on the front page of the Association website. Manufacturers are identified by the letter "M" and their agents by the letter "A."

The other important factor is to make sure that the equipment has either been CE marked or carries an FCC number (cf Chapter 15: Regulation). This, at least, shows that the manufacturer is aware of the current licensing regulations and abiding by them. Never, ever use or buy GPR equipment which breaks the existing regulatory framework. This could potentially result in being prosecuted for illegally transmitting. If a transmission breach occurs, the GPR operator using unlicensed equipment would be in a very vulnerable position legally.

There have also been cases of nonlicensed equipment failing to perform as it should and there is very little that anyone can do about this beyond reporting the seller to the local Trading Standards Association.

Unfortunately, in spite of the wide availability of GPRs, there has been and continues to be a low level of problems in sourcing reliable

GPR systems. The majority of GPR manufacturers are responsible and professional and produce equipment which is both fit for purpose and also meets the regulations appropriate to the countries in which their equipment is sold. They are generally also capable of advising what the current regulations are. However, there are two major problems which it is important for newcomers to the technology to understand and be wary of before either beginning surveying or contemplating the purchase of a GPR, namely overzealous marketing and unusual technical claims. It is also important for those commissioning GPR surveys to be aware of potential equipment problems because these may affect their supplier of GPR services and information.

OVER ZEALOUS MARKETING

One common problem is overly enthusiastic selling by an agent of a responsible manufacturer who does not fully understand either the underlying physical principles or the licensing requirements for all GPRs. Claims in this category include, for example, confusion over the relevance of the central frequency. Central frequency is the frequency transmitted at the peak energy of the antenna in use (cf Chapter 2: Wavelengths and Why They Matter). Each antenna has its own frequency range around this central frequency. There is no general agreement as to the limits of the range of each antenna because this depends on a technical judgment as to where the energy level is so low as to be negligible. It is difficult to reach a general agreement among manufacturers who are, after all, in competition with each other. However, $0.5x$ to $2x$ where x is the central frequency is a reasonable prescription of the frequency range, independent of manufacturer. A claim, therefore, that there is no need to use a 250 MHz antenna because the 500 MHz antenna "covers that frequency anyway" is extremely ill-judged. The 500 MHz antenna will cover the range from 250 MHz to 1 GHz but the energy at the lower end could never match or even approach the peak energy of the lower frequency antenna. There is no way in which this antenna could match the probing depth of the 250 MHz antenna whose own range, based on the same formula, would be from 125 to 500 MHz. There is nothing wrong with the equipment in this example but the person offering it has not understood why frequency matters and why his/her client requires the lower frequency antenna.

Into this category also falls the claim that a 250 MHz antenna can reach 8 m of depth penetration, even in wet clay. Here there are a number of major misunderstandings in respect of the effect of water on the wavelength, and the potential attenuation of wet clay. First of all for the depth probed by the radar to reach 8 m, the wavelength will have to be in excess of 0.4 m. (Assuming a perfect, nonattenuating soil in which 20 wavelengths of depth penetration can be achieved, gives a wavelength of 8/20 or 0.4 m). Secondly clays, of whatever variety, are well known not to be perfect nonattenuating soils so the wavelength will have to be considerably longer than this estimate. For a 250 MHz antenna to produce a wavelength of 0.4 m, the transmission velocity would have to be 0.1 m/ns which would normally represent a very dry soil. (This is based on the formula $\lambda = v/f$. $V = 0.4*0.25$ or 0.1 m/ns). Lastly wet clay contains the combination of free ions and mobility which means that the likelihood is that much, if not all, of the signal would be lost as a weak electric signal well before 8 m depth could be reached (cf Chapter 4: The Effect of Water and Air). The claim is therefore completely unrealistic on a number of counts.

This also demonstrates that just being able to calculate wavelengths is potentially sufficient to protect the unwary from this type of claim. The wavelength emitted by a 250 MHz antenna in a dry soil is 0.4 m (taking 0.1 m/ns as the transmission velocity in a dry soil). Since water slows down transmission, effectively reducing the wavelength, the claim is immediately seen as ridiculous, even before considering the effects of attenuation.

In the personal experience of the author, this type of problem is relatively more common in Asia than in Europe or America but it can still be found worldwide. One possible cause of the problem is a desire to supply equipment but not having the appropriate parts to hand and being unwilling to go to the expense of bringing in more antennas. Another common motivator is the desire to impress the client with the superiority of the system on offer over other equivalent systems but not having sufficient technical understanding of how GPRs work to realize that they are demonstrating extreme naivety rather than the technical superiority of their equipment. The solution is, of course, education!

EuroGPR MEMBERSHIP

The European GPR Association has a number of functions. It represents the GPR industry at European level over licensing issues (cf Chapter 15: Regulation). It develops good practice guidelines for its members for the various applications of GPR technology. It hosts a virtual library via its website which can provide suitable sources of information and some technical papers and it is in the course of developing an online training course. Above all, however, the Association is attempting to demonstrate and hold to best possible practice in the use of a technology which is not intuitive and therefore often misunderstood.

A selection of members is listed on the front page of the website in their capacities as manufacturer, agent, service provider, or researcher. This can be expanded into a list of all members. In addition, all members hold a certificate of membership with a unique number and also the year of validity. It is therefore possible to verify the current status of any membership.

Applications to join are not accepted automatically. Applicants of all categories are expected to provide good references from two referees who are familiar with their GPR work in support of their application for membership. Potential members have to accept both the Code of Ethics and the Memorandum of Association. Although the Association is predominantly European in its base, applications are accepted from countries outside Europe. Current membership includes several members based in other continents.

The information submitted by the applicant is then reviewed by the Association committee. Acceptance may be unconditional, for example, for experienced GPR users with good references and perhaps membership of other recognized reputable organizations. For less experienced potential members or those beginning to use GPR, conditional acceptance may be given. This will become unconditional when they have had a chance to prove themselves as competent and reliable. Once new members have been accepted, they are also required to make a short presentation of their GPR work to their fellow members at the next meeting.

This is not to say that membership of EuroGPR is the only proof of professional competence. As outlined in Chapter 15, Regulation (other organizations), there are a number of professional organizations for which membership implies both competence and best practice.

UNUSUAL TECHNICAL CLAIMS

The first problem is that because of the physical limitations of GPR, there are equipment producers and/or service providers who claim to be able to circumvent these restrictions, i.e., in essence to be able to overcome the basic laws of Physics which underlie the operation of all radars, including GPR. For example, as we saw in Chapter 2, Wavelengths and Why They Matter, and Chapter 7, Survey Strategies, it is not possible to use one single radar for all depths. The maximum depth which can be reached is a function of the wavelength of the radar and the lossiness of the soil (or other survey medium). A common inappropriate claim is to be able to reach much greater depths than with a "traditional" GPR either irrespective of the frequency of the equipment or giving no details of frequency.

Another example is a claim to be able to detect any size of object, regardless of the depth at which it lies. Again, this is totally inappropriate. The ability to detect depends on the size of the object being a reasonable proportion of the transmitted wavelength. If the object is too small, it will not be detected. A small object which is deeply buried will either not be detectable because the wavelength to reach that depth is too large for the size of the object or, if the wavelength is short enough to detect the target, it is likely that the electromagnetic pulses will be unable to reach the depth. A notorious example of this is given in the preface to Daniels (2004).

A third example of an inappropriate claim is that a particular radar can detect in very lossy conditions, e.g., saltwater. The technology commonly used at sea, for example, is sonar, not radar. As seen in Chapter 7, Survey Strategies, it is possible to operate a radar in freshwater but not where there is freely mobile ion content. The limiting factor for freshwater is the slower transmission velocity which effectively decreases the wavelength. In saline water, the signals pass into the solution as a weak electric current and are not returned to the receiver antenna. The result is ringing, repeated echo effects from the surface. Even dry sand with salt content can be an extremely difficult survey medium for GPR.

Note that these types of claim generally rely on the user of the equipment and/or the person commissioning the survey information not understanding the basics of how GPR works.

The motivation for this type of claim is obvious. Radar is a very powerful tool. However it would be much easier to use if it were not necessary to work out which frequency was appropriate for the investigation and if the size of target and/or the depth of the target did not matter. Equally being able to ignore the signal losses (attenuation) caused by soil or water (with or without saline content) would be an advantage. Unfortunately, however, this is not how radar works and this type of claim needs to be treated with caution.

Since the question of the reliability of this type of claim comes down to the fundamentals of the underlying Physics, there is a simple solution to the problem. Where a claim is reliable, it should be possible to demonstrate this in a blind test, i.e., where the operator has no access to any subsurface information so that any claims made are capable of being independently verified. If it is not possible, for whatever reason, to complete a blind test, then the claims made regarding the equipment should be treated with caution, being potentially literally "too good to be true."

This is not to say that there can never be improvements in equipment performance. All reputable manufacturers make regular claims in this regard. However they do not ignore the underlying physical principles, i.e., the essentials of transmission and reception of radio waves. The improvements are never as outlandish or as extreme as the examples given here. The equipment designers are also usually quite prepared to substantiate their technical claims by demonstration, including blind testing, and by publishing in mainstream GPR meetings and conferences or in peer-reviewed journals where they accept potential challenges from technical experts. Although they will not normally reveal all the details of how they have achieved their advances, they will give sufficient information to demonstrate that they conform to widely accepted Physical principles, rather than "improving" on them. They do not normally hide behind claims that a blind test would unfairly advantage their commercial competitors (for which patenting would, in any case, provide protection).

One minor variation on the above problem which is sometimes also encountered is equipment which is actually a different type of device such as a metal detector or a magnetometer being marketed as a GPR. Metal detectors and magnetometers are two examples of valid geophysical tools but they are not GPRs and do not function in the same manner. They need to be judged against their own equivalents.

There is some guidance available on the subject which could prove useful in forming a personal judgment: see the "Technology You Can Trust In" section of the EuroGPR website for further information.

REFERENCES

EuroGPR website. <www.eurogpr.org>.

Daniels, D.J., 2004. Ground Penetrating Radar, second ed. The Institution of Electrical Engineers, London, UK, p. xvi (Preface).

EuroGPR website. http://www.eurogpr.org/vn2/index.php/2-uncategorised/9-technology-you-can-trust-in is the direct reference for Technology You Can Trust In on the EuroGPR website. Alternatively, there is a link from the Home page.

Index

Note: Page numbers followed by "*f*" and "*t*" refer to figures and tables, respectively.

Printed in the United States
By Bookmasters

Printed in the United States
By Bookmasters